Teen Guides to

Environmental Science

Teen Guides to
Environmental Science

People and Their Environments
VOLUME III

John Mongillo
with assistance from Peter Mongillo

Greenwood Press
Westport, Connecticut • London

Library of Congress Cataloging-in-Publication Data

Mongillo, John F.
  Teen guides to environmental science / John Mongillo with assistance from Peter Mongillo.
    p. cm.
  Includes bibliographical references and index.
    Contents: v. 1. Earth systems and ecology—v. 2. Resources and energy—v. 3. People
  and their environments—v. 4. Human impact on the environment—v. 5. Creating a
  sustainable society.
  ISBN 0–313–32183–3 (set : alk. paper)—ISBN 0–313–32184–1 (v. 1 : alk. paper)—
  ISBN 0–313–32185–X (v. 2 : alk. paper)—ISBN 0–313–32186–8 (v. 3 : alk. paper)—
  ISBN 0–313–32187–6 (v. 4 : alk. paper)—ISBN 0–313–32188–4 (v. 5 : alk. paper)
    1. Environmental sciences. 2. Human ecology. 3. Nature–Effect of human beings on. I.
  Mongillo, Peter A. II. Title.
  GE105.M66   2004
  333.72—dc22   2004044869

British Library Cataloguing in Publication Data is available.

Library of Congress Catalog Card Number: 2004044869
ISBN: 0–313–32183–3 (set)
        0–313–32184–1 (vol. I)
        0–313–32185–X (vol. II)
        0–313–32186–8 (vol. III)
        0–313–32187–6 (vol. IV)
        0–313–32188–4 (vol. V)

First published in 2004

Greenwood Press, 88 Post Road West, Westport, CT 06881
An imprint of Greenwood Publishing Group, Inc.
www.greenwood.com

Printed in the United States of America

The paper used in this book complies with the
Permanent Paper Standard issued by the National
Information Standards Organization (Z39.48–1984).

10 9 8 7 6 5 4 3 2 1

CONTENTS

CHAPTER 3 The Agricultural Revolution: Expansion of Agricultural Productivity 33

CHAPTER **4** The Industrial Revolution: Economic, Social, and Environmental Changes　　50

CHAPTER **5** Economic Expansion and the Environment　　73

CHAPTER **6** The Growing Human Population: Food Supply and Social Issues · 87

ACKNOWLEDGMENTS

The authors wish to acknowledge and express the contribution of the many nongovernment organizations, corporations, colleges, and government agencies that provided assistance to the authors in the research for this book. The authors are grateful to the Greenwood Publishing Group for permission to excerpt text and photos from *Encyclopedia of Environmental Science*, John Mongillo and Linda Zierdt-Warshaw, and *Environmental Activists*, John Mongillo and Bibi Booth. Both books are excellent references for researching environmental topics and gathering information about environmental activists. Many thanks to those who provided special assistance in reviewing particular topics and offering comments and suggestions: Sara Jones, middle school director for La Jolla Country Day School in San Diego, California; Emily White, teacher of geography and world cultures at the 5th grade level at La Jolla Country Day School, San Diego, California; Lucinda Kramer and John Guido, middle school social studies coordinators, North Haven, Connecticut; Daniel Lanier, environmental professional, and Susan Santone, executive director of Creative Change, Ypsilanti, Michigan.

A special thank you goes to the following people and organizations that provided technical expertise and/or resources for photos and data: Neil Dahlstrom, John Deere & Company; Francine Murphy-Brillon, Slater Mill Historic Site; Lake Worth Public Library, Florida; Pacific Gas & Electric; Energetch; Environmental Justice Resource Center; NASA Johnson Space Center; Seattle Audubon Society; John Onuska, INMETCO; Cathrine Sneed, Garden Project; Denis Hayes, president, Bullitt Foundation; Ocean Robbins, Youth for Environmental Sanity; Maria Perez and Nevada Dove, Friends of McKinley; Juana Beatriz Gutiérrez, cofounder and president of Madres del Este de Los Angeles—Santa Isabel; Mikhail Davis, director, Brower Fund, Earth Island Institute; Randall Hayes, president, Rainforest Action Network; Tom Repine, West Virginia Geologic Survey; Peter Wright and Nancy Trautmann, Cornell University; Mary N. Harrison, University of Florida; and Huanmin Lu, University of Texas, El Paso.

Other sources include Centers for Disease Control and Prevention, Department of Environmental Management, Rhode Island; ChryslerDaimler; Pattonville High School; National Oceanic and

Atmospheric Administration; Chuck Meyers, Office of Surface Mining; U.S. Department of Agriculture; U.S. Fish and Wildlife Service; U.S. Department of Energy; U.S. Environmental Protection Agency; U.S. National Park Service; National Renewable Energy Laboratory; Tower Tech, Inc.; Earthday 2000; Marilyn Nemzer, Geothermal Education Office; U.S. Agricultural Research Service; U.S. Geological Survey; Glacier National Park; Monsanto; CREST Organization; Shirley Briggs, Vortec Corporation; National Interagency Fire Center/Bureau of Land Management; Susan Snyder, Marine Spill Response Corporation; Lisa Bousquet, Roger Williams Park Zoo, Rhode Island; Netzin Gerald Steklis, International National Response Corporation; U.S. Department of the Interior/Bureau of Reclamation; Bluestone Energy Services; OSG Ship Management, Inc.; and Sweetwater Technology.

In addition, the authors wish to thank Hollis Burkhart and Janet Heffernan for their copyediting and proofreading support; Muriel Cawthorn, Hollis Burkhart, and Liz Kincaid for their assistance in photo research; and illustrators Christine Murphy, Susan Stone, and Kurt Van Dexter.

The responsibility of the accuracy of the terms is solely that of the authors. If errors are noticed, please address them to the authors so that corrections can be made in future revisions.

INTRODUCTION

Teen Guides to Environmental Science is a reference tool which introduces environmental science topics to middle and high school students. The five-volume series presents environmental, social, and economic topics to assist the reader in developing an understanding of how human activity has changed and continues to change the face of the world around us.

Events affecting the environment are reported daily in magazines, newspapers, periodicals, newsletters, radio, and television, and on Websites. Each day there are environmental reports about collapsing fish stocks, massive wastes of natural resources and energy, soil erosion, deteriorating rangelands, loss of forests, and air and water pollution. At times, the degradation of the environment has led to issues of poverty, malnutrition, disease, and social and economic inequalities throughout the world. Human demands on the natural environment are placing more and more pressure on Earth's ecosystems and its natural resources.

The challenge in this century will be to reverse the exploitation of Earth's resources and to improve social and economic systems. Meeting these goals will require the participation and commitment of businesses, government agencies, nongovernment organizations, and individuals. The major task will be to begin a long-term environmental strategy that will ensure a more sustainable society.

CREATING A SUSTAINABLE SOCIETY

Sustainable development is a strategy that meets the needs of the present without compromising the ability of future generations to meet their own needs. Many experts believe that for too long, social, economic, and environmental issues were addressed separately without regard to each other. In creating a sustainable society, there needs to be an integration of goals related to economic growth, environmental protection, and social equity. Some of these integrated sustainable goals include the following:

- Improve the quality of human life

- Conserve Earth's diversity

- Minimize the depletion of nonrenewable resources

- Keep within Earth's carrying capacity

- Enable communities to care for their own environments

- Integrate the environment, economy, and human health into decision making

- Promote caretakers of Earth.

OVERVIEW

Teen Guides to Environmental Science provides an excellent opportunity for students to study and focus on the integration of ecological, economical, and social goals in creating a sustainable society. Within the five-volume series, students can research topics from a long list of contemporary environmental issues ranging from alternative fuels and acid rain to wetlands and zoos. Strategies and solutions to solve environmental issues are presented, too. Such topics include soil conservation programs, alternative energy sources, international laws to preserve wildlife, recycling and source reduction in the production of goods, and legislation to reduce air and water pollution, just to name a few.

Major Highlights

- Assists students in developing an understanding of their global environment and how the human population and its technologies have affected Earth and its ecology.

- Provides an interdisciplinary perspective that includes ecology, geography, biology, human culture, geology, physics, chemistry, history, and economics.

- "Raises a student's awareness of a strategy called sustainable development that meets the needs of the present without compromising the ability of future generations to meet their own needs" (Bruntland Commission). The strategy includes a level of economic development that can be sustained in the future while protecting and conserving natural resources with minimum damage to the environment. People concerned about sustainable development suggest that meeting the needs of the future depends on how well we balance social, economic, and environmental objectives—or needs—when making decisions today.

- Presents current environment, social, and economic issues and solutions for preserving wildlife species, rebuilding fish stocks, designing strategies to control sprawl and traffic congestion, and developing hydrogen fuel cells as a future energy source.

- Challenges everyone to become more active in their home, community, and school in addressing environmental problems and discussing strategies to solve them.

ORGANIZATION

Teen Guides to Environmental Science is divided into five volumes.

Earth Systems and Ecology

Volume I begins the discussion of Earth as a system and focuses on ecology—the foundation of environmental science. The major chapters examine ecosystems, populations, communities, and biomes.

Resources and Energy

Currently, fossil fuels drive the economy in much of the world. In Volume II conventional fuels such as petroleum, coal, and natural gas are reported. Other chapters elaborate on nuclear energy, hydrogen energy, wind energy, geothermal energy, solar energy, and natural resources such as soil and minerals, forests, water resources, and wildlife preserves.

People and Their Environments

The history of civilizations, human ecology, and how early and modern societies have interacted with the environment is presented in Volume III. The major chapters highlight the Agricultural Revolution, the Industrial Revolution, global populations, and economic and social systems.

Human Impact on the Environment

Volume IV discusses the causes and the harmful effects of air and water pollution and sustainable solution strategies to control the problems. Other chapters examine the human impact on natural resources and wildlife and discuss efforts to preserve them.

Creating a Sustainable Society

Volume V focuses on the importance of living in a sustainable society in which generations after generations do not deplete the natural resources or produce excessive pollutants. The chapters present an overview of sustainability in producing products, preserving wildlife habitats, developing sustainable communities and transportation systems, and encouraging sustainable management practices in agriculture and commercial fishing. The last chapter in this volume considers the importance of individual activism in identifying and solving environmental problems in one's community.

PROGRAM RESEARCH

The five-volume series represents research from a variety of recurring and up-to-date sources, including newspapers, middle school and high school textbooks, trade books, television reports, professional journals, national and international government organizations, nonprofit organizations, private companies, businesses, and individual contacts.

CONTENT STANDARDS

The series provides a close alignment with the fundamental principles developed and reported in the President's Council on Sustainable Development and the learning outcomes for middle school education standards found in the North American Association for Environmental Education, the National Geography Standards, and the National Science Education Standards.

MAJOR ENVIRONMENTAL TOPICS

The *Teen Guides to Environmental Science* provide terms, topics, and subjects covered in most middle school and high schools environmental science courses. These major topics of environmental science include, but are not limited to:

- Agriculture, crop production, and pest control
- Atmosphere and air pollution
- Ecological economies
- Ecology and ecosystems
- Endangered and threatened wildlife species
- Energy and mineral resources
- Environmental laws, regulations, and ethics
- Oceans and wetlands
- Nonhazardous and hazardous wastes
- Water resources and pollution.

SPECIAL FEATURES

Tables, Figures, and Maps

Hundreds of photos, tables, maps, and figures are ideal visual learning strategies used to enhance the text and provide additional information to the reader.

Vocabulary

The vocabulary list at the end of each chapter provides a definition for a term used within the chapter with which a reader might be unfamiliar.

Marginal Topics

Each chapter contains marginal features which supplement and enrich the main topic covered in the chapter.

Activities

More than 100 suggested student research activities appear at the ends of the chapters in the books.

In-Text References

Many of the chapters have specially marked callouts within the text which refer the reader to other books in the series for additional information. For example, fossil fuels are discussed in Volume V; however, an in-text reference refers the reader to Volume II for more information about the topic.

Websites

A listing of Websites of government and nongovernment organizations is available at the end of each chapter allowing students to research topics on the Internet.

Bibliography

Book titles and articles relating to the subject area of each chapter are presented at the end of each chapter for additional research opportunities.

Appendixes

Four appendixes are included at the end of each volume:

- Environmental Timeline, 1620–2004. To understand the history of the environmental movement, each book provides a comprehensive timeline that presents a general overview of activists, important laws and regulations, special events, and other environmental highlights over a period of more than 400 years.

- Endangered List of U.S. Wildlife Species by State.

- Website addresses by classification.

- Government and nongovernment environmental organizations.

Prehistoric Human Societies in the Stone Age

Each day, human demands for food, water, energy, shelter, and employment alters the natural environment. The human impact on vegetation, soil, water, and mineral resources has caused pollution, loss of wildlife and habitats, and soil erosion.

Exactly when and how did human societies begin altering their environment? This chapter traces the early development of prehistoric human societies and their interaction with Earth's environment. The text will describe the kinds of natural resources, technology, and energy used by these prehistoric human societies and what the impact, if any, these early societies had on their environment.

The Stone Age can be divided into two major time periods: the Old Stone Age and the New Stone Age. The Old Stone Age was a time when hominids used crudely made tools. The New Stone Age was a period when communities used more refined and polished tools.

PALEOLITHIC PERIOD

According to archeologists, who study the life and culture of early people and places, the early human ancestors, called *hominids*, date back to about 2.5 million years ago. The hominids formed social organizations and lived in small communities or tribes. Known as hunter-gatherers, they were efficient predators who hunted and fished for food. They also gathered nuts, berries, seeds, wild fruits, and vegetables to supplement their diet. They found food in a variety of biomes such as grasslands, lakes, and forests.

Hunting and Gathering Societies

Hunter-gatherers were nomadic tribes comprising between 70 and 150 people in each group. The tasks in the tribes involved a division of family labor between men and women. Some members in the group did the hunting while others cared for the young in the family and prepared the meals.

The roving bands were on the go much of the year and did not settle in any particular place for any length of time. They probably lived most of their time outdoors, or they found shelter in caves or made

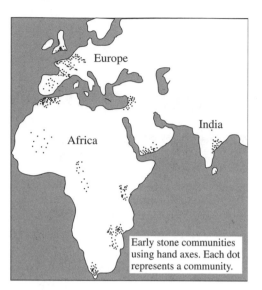

FIGURE 1-1 • Early Stone Tool Communities

Early stone communities using hand axes. Each dot represents a community.

temporary lean-tos. They did not farm or herd animals. They had limited *technology*—mostly stone tools. However, it was the development of these tools that enabled them to dig up roots to eat, cut up meat and fish, chop wood, sew clothes, make fires, and fend off predators.

Much of what we know about the hunter-gatherers comes from the tools they left behind during a time called the Stone Age. The term Stone Age, used by historians and archaeologists since the early 1800s, refers to an earlier, prehistoric stage of human society, a society in which stone rather than metal tools were used. Stone was probably the most important mineral resource of the early people. Wood was also used, however, for digging and building lean-tos and ladders.

The dates of the Stone Age vary considerably for different parts of the world. In Europe, Asia, and Africa, the Stone Age began between 1.75 and 2.5 million years ago. The Stone Age in North and South America began when prehistoric humans migrated from other lands into the New World, some 30,000 years ago. Estimations of the prehistoric world population of hominids during the Stone Age period ranges between 600,000 and 1,000,000 inhabitants.

Old Stone Age

At the beginning of the Pale Age, many of the stone tools were just large round rocks with pieces chipped out of them. For the most part, a crude stone tool consisted of a rock chipped away to form a cutting edge at one end. Crude stone tools have been found with the remains of prehistoric people in various places in Africa, primarily northern Kenya, southern Ethiopia, and Tanzania. The stone tools found in these areas date from 1.75 million to about 2.5 million years ago. Stone tools have been found in other places in the world—Asia, the Middle East, Europe, and North and South America—but these tools are not as old as the ones found in Africa.

Refer to Volume II for more information about indigenous peoples.

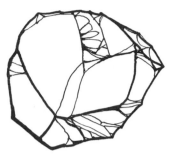

FIGURE 1-2 • Early ancient tools were usually simple rocks whose surfaces were chipped away on all sides. These rough, sharp-edged tools were probably used to break up meat on bones and as a throwing stone for hunting.

Indigenous People

In many ways, the *indigenous* people living today remind us about how early human societies lived. Indigenous people, or those who are native to their environment, number about 600 million worldwide today, and they live in many areas of the world. Most places are remote and forested and have fragile ecosystems. Anthropologists believe that the largest number of indigenous people live in the Amazon region of South America.

Some of the indigenous groups who live in forests throughout the world include the Yanomami in South America, the Aka in southwestern Africa, the Ifugao in the Philippines, and the Iban in western Borneo. Some of these tribes lead a nomadic life hunting and gathering while others hunt and fish too, but they farm as well, and they trade forest products with their neighbors.

Each indigenous tribe has its own traditions, social values, language, and culture. Many live in small, scattered villages and few, if any, own land or have many possessions. Environmentalists call many of these groups conservationists who use *sustainable* practices. These tribes take only those resources that are needed and conserve resources in their area to protect them for future generations. Currently a large number of indigenous tribes are threatened by the encroachment of the activities of modern human society, such as ranching, farming, and road building.

Stone Age Technology

Flint was the most desirable kind of stone for Stone Age toolmakers. In fact, it was so valuable that flint was traded between tribes that had the stone with those who did not.

In places where flint was unobtainable, quartz and other rocks were used. Obsidian, a volcanic glasslike rock which is gray to black and nearly *opaque*, can be extracted from layers of limestone and chalk. Quartzite was also used in toolmaking, particularly in India. Generally, however, most rock tools with edges were made of flint.

FLINT TOOLS AND FIRES

Flint is harder than most steel. When broken, flint fractures in such a way that it leaves a sharp edge on the broken pieces. The pieces are sharp enough to cut through animal skins and to cut branches and twigs. Flint was also used by prehistoric people to start fires. Striking

The Penan hunters are indigenous people who live in the tropical rainforest of present day South America. Indigenous people are those who are native to their environment. (Courtesy of Rainforest Action Network)

Some ancient hunters utilized quartz, a common mineral, as a stone tool. However, flint was the most abundant material used during the Stone Age. (Courtesy of United States Department of Agriculture. Photo by Ken Hammond.)

DID YOU KNOW?

When glass breaks, it fractures into pieces similar to the way in which flint fractures when made into tools. The broken, brittle pieces of glass are very sharp, which is why one must be careful in picking up broken glass.

two flint stones together produces sparks which ignite fires in dry grass or wood shavings. Making fire was important for cooking and for providing heat during colder temperatures. Special stone hearths were built for cooking. Hunters also used fire to burn fields of dry grass to encircle the migrating animal herds. Once the animals were cornered, the hunters moved in on those who could not escape the ring of fire.

Putting Stone Tools to Work

Most of the rocks used by the toolmakers were found in nearby streams. After a rock had been collected, it was chipped away until it became the desired shape of a tool. Another method used by the toolmaker, called percussion flaking involved striking a portion of the rock with a stone hammer. The left over thin pieces or flakes that were chipped off the rock became tools with sharp edges suitable for cutting meat or hides.

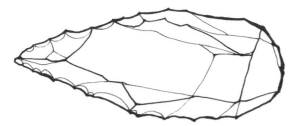

FIGURE 1-3 • Hand axe technology came in all kinds of shapes. The hand axe had a variety of uses that may have included such tasks as skinning and cutting up meat and hacking wood. The axe was probably used also as a tool to dig up roots and tubers.

Stonehenge is a prehistoric stone monument located in England. The monument dates back to more than 3,500 years ago and may have been used as a ceremonial center. Stone hammers were used to shape much of the blue stone used in the monument. (Courtesy of James and Patricia Mongillo)

HAND AXES

One of the handiest of all stone tools was the hand axe. The hand axe was a rock about the size of a baseball that had a sharp point. The hand axe could be used for a variety of jobs. It was probably used mostly to dig up edible roots, tubers, and other plants buried deep in the ground. Hand axes could also be used for cutting meat, scraping skins, chopping branches and twigs, digging holes, and hammering wood. It was probably used to fend off wild animals as well.

Hand axes come in many shapes and sizes, and many styles unique to cultures of specific periods and in specific geographical areas. Almost all hand axes have a point and were shaped to be held. Very few hand axes had notches for mounting. Most hand axes were about 10 centimeters (4 inches) long; however, some hand axes were more than 15 centimeters (about 6 inches) long and about 7 centimeters (2.5 inches) wide.

Stone tools came in a variety of shapes and sizes. One kind of stone tool was in the shape of a chisel, which was used to smash and

DID YOU KNOW?

Like humans, chimpanzees are tool-makers and tool users. Jane Goodall, a chimpanzee science researcher, has recorded chimps making and using simple tools such as sticks and rocks for digging and cracking nuts. They also chew up leaves and use them as sponges for wiping up water. Some chimps are even known to carry tools around with them.

crack bones into splinters. The splinters were made into needles with eyes. Other stone tools were used as awls; these had very sharp points and were used to punch holes into animal skins. Using the awl and the eyed needle made it possible for humans to stitch and sew skins and fur into clothing, headgear, blankets, and footwear. Having proper warm clothing allowed some groups to migrate and live in colder regions of the world.

Hand axes were also used to construct huts and other dwellings. One kind of hut was covered with cut-up pieces of animal hides weighed down on the sides with bones. The huts were made snug with skins and turf coverings.

Location of Stone Tools

INDIA

During the 1860s, archaeologists discovered a site with a large number of Stone Age tools. The site, in Tamil Nadu state in southern India, was not much studied or written about. In 1991 new research and digging at the same site revealed fragments of stone tools dating back to 1 million years ago. The stone tools were the first ever found in clay deposits anywhere in India. The tools included hand axes, cleavers, picks, awls, scrapers, knives, and stone flakes. The tools were made of quartzite, a local rock.

MALAYSIA

Archaeologists have found ancient stones and tools, believed to be more than 100,000 years old, in a remote village in Malaysia. Among the items found in several trenches dug in the area are tools for cutting meat, chopping wood, and making other tools out of bamboo and wood. The people who lived in the area at that time were nomads who hunted aquatic and jungle animals and gathered fruits and vegetables.

Cultural Development: Stone Age Art

Rock shelters and caves in many parts of India are host to a wealth of beautiful and ancient paintings. Early hunter-gatherers painted animals, birds, plants, and scenes of hunting and gathering on the walls and roofs of their caves. They used mineral dyes and painting tools to sketch the animals, plants, and scenes. The earliest paintings date to about 8,000 years ago.

The most famous rock art images come from the caves and rock shelters of central India.

Rock paintings have been also been discovered in caves in Spain, France, and Africa. In France, cave animal paintings date back as much as 25,000 years ago. The paintings depict woolly mammoths, fish, horse, and bisons. One painting shows a mammoth caught in a trap. Other paintings illustrate figures who are part human and part animal. Prehistoric people also made sculptures and stone engravings.

OTHER PLACES

Early people in Ireland and Finland used tools and weapons made from stone and wood. They lived almost entirely by hunting and fishing. Their clothes were probably made of animal hides. The Stone Age people in Finland often established their sites at shore locations in open, sandy areas between the water and adjoining forest. Hand axes, about 800,000 years old, were found in southeast China documenting that prehistoric people in Asia were also skilled in shaping tools.

The prehistoric peoples of Alberta, Canada, used a variety of raw materials to make tools for hunting, scraping hides, and preparing food. The toolmakers in Alberta used mostly chert and quartzite, both of which could be found along riverbeds. They also used some obsidian to make some of their tools.

Refer to Volume I for more information about rocks and minerals.

New Stone Age Technology

The New Stone Age, from about 10,000 to 12,000 years ago, was highlighted by a new kind of toolmaker who was producing more streamlined, more effective weapons and tools. As a result of the new tools, hunter-gatherers could exploit more animal and plant resources. Their hunting may also have begun the extermination of some wildlife species.

The toolmaker had to perform additional shaping and thinning to make these kinds of tools, which included arrowheads or spear points. For a spear point, for example, a larger flake was selected and shaped by using a stone hammer to strike off several broad, thin flakes from both surfaces. Once the shape was determined, the final stage of thinning the point and sharpening the edges began. This process was

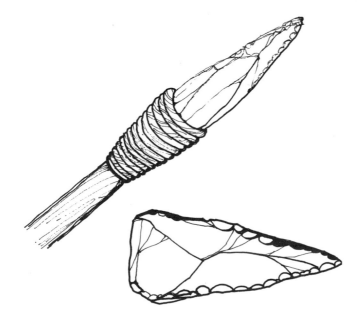

FIGURE 1-4 • Very sharp-pointed stones, usually made from flint, were mounted on long sticks that were used for spearing animals and self-defense. Many of these tools were found at certain Cro-Magnon sites.

FIGURE 1-5 • Early Stone Age communities made remarkable technological changes working with stone and bone materials. Jewelry-like bone ornaments have been found in archeological excavations in Europe; some of the necklaces recovered were made from animal parts. The illustration atop shows the design of a bone necklace made from the pointed teeth of animals such as wolves. The illustration of the necklace at the bottom was made from shells.

known as pressure flaking. Here, small, thin flakes were pushed off the edges of the tool using an antler or bone punch. Then the tool was polished by using fine sand and water. The toolmaker had to use both skill and physical strength to make these new kinds of stone tools. Tens of thousands of these finely made arrowhead and spear points have been found in sites as old as 10,000 years throughout much of world.

BONE TOOLS AND ORNAMENTS

During the New Stone Age, many of the animals that were used for food and clothing also supplied a ready source of raw materials for tools and ornaments. Hard stone tools, such as serrated blades and small scrapers with chisel-like ends were used to make bone tools and ornaments. Bone was used to make barbed fishhooks, arrowheads, eyed needles, and small leather-working awls. Spines from some animals were sharpened to points for use as fishhooks and needles. Some bone tools were polished to a high gloss with a fine-grained grinding stone. Occasionally, these tools were decorated with finely marked lines.

Many kinds of animal bones were used to make ornaments, including rings, bracelets, hairpins, necklaces, beads, and pendants. Holes were drilled through tiny animal bones to make beads for necklaces. Necklaces were often made from the pointed teeth of various animals. Hairpins were made from deer bones.

DID YOU KNOW?

Early hominids in South Africa used sticklike tools to dig into termite mounds. Termites and other insects were important foods in early human diets. Sticklike bone tools were also used to dig up tubers and other underground plants.

DID YOU KNOW?

The first known musical instrument is a flutelike instrument made of bone. Discovered in North Africa, the instrument is 47,000 years old.

NEOLITHIC PERIOD AND GATHERING SOCIETIES DECLINE

Towards the end of the Stone Age (about 10,000 B.C.), a new chapter of human history culture and technology began to emerge. Archeologists refer to this period as the *Neolithic Revolution*, a period when human societies abandoned their nomadic existence. Although some hunter-gatherer groups still roamed about, other groups began to settle down in one place. Human societies began to farm and herd animals. In time, the hunting-gathering societies began to decline further. In its place, other societies started to grow their own plants for food and to tame and herd animals for meat and milk products and hides for clothing. These activities began to put more demand on Earth's resources.

Cultivation of Wild Plants

Farming or agricultural activities first took place in the Middle East, northeast Africa, Europe, and South America. Farming probably began when people, who had been gathering plants, such as wild wheat, observed that where they dropped seeds, new plants grew. As a result, they began dispersing seeds and planting gardens where they planned to stay for a period of time. Once the plants were harvested, some of the left over seeds were saved to grow the next crop. After several harvests, the farmers concluded that the best seeds produced the best crops, and they began to select and save the seeds with the most desirable characteristics for cultivation. The *domestication* of wheat from wild plants was developed from about 6,000 to 7,000 years ago.

The early farmers grew different species of wild wheat called einkorn and emmer. Together with barley, the wheats were the major grains of the Neolithic Period. Cereal crops and vegetables, as well as

TABLE 1-1 **Domesticated Plants**

Area	Plants	Date of Domestication (estimate)
1. Southwest Asia	Wheat, pea, olive	8500 B.C.
2. China	Rice, millet	by 7500 B.C.
3. South America	Corn, beans, squash, potato, manioc	by 3500 B.C.
4. Eastern United States	Sunflower, goosefoot	2500 B.C.
5. Africa	Sorghum, African rice, African yam, oil palm, coffee, sugar cane, banana	7000 B.C. to 3000 B.C.

Stone Age Diet

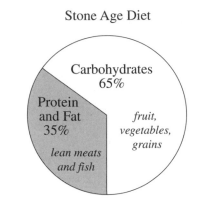

FIGURE 1-6 • Stone Age Diets Early humans obtained more than half of their calories from carbohydrates. They ate 35 percent of their calories in protein from lean wild animals including fish. They ate more than 5 times today's level of fiber, which was about 100 grams a day.

DID YOU KNOW?

Thousands of years ago, Neolithic farmers grew wheat. Today, wheat ranks first in the world's grain production and accounts for more than 20 percent of the total calories consumed by humans.

fruits, were widely grown in different parts of the world. As an example, farmers in China grew millet and rice, while those living in the Middle East grew onions, peas, lentils, and dates, as well as barley and wheat.

Farmers in southern Europe grew grapes, olives, oats, and rye grains. Cotton was domesticated in northern Pakistan. In the Americas, farmers grew potatoes, pumpkins, squash, and beans. Farming produced more food than the older methods of hunting and gathering. Farming allowed people to stay in one place for more than a season. They could build permanent shelters. They could store food to tide them over during times of droughts and cold seasons. The food surpluses relieved some people from farm tasks to become merchants and traders.

Domestication of Animals

Animals were of vital importance to the farmers of the Neolithic Period. The energy of the animals allowed a farmer to grow more food than the animal ate. Animals provided clothing and food, and some were used for protection. Land management became easier and tasks quicker. On the other hand, animals had to be fed, which meant that the farmers had to grow food for them as well as for their families.

Neolithic people discovered that some animals could be tamed and kept in pens. The domestication of sheep and goats was a good choice because these animals were able to survive all kinds of outdoor climates and live on poor land not used for farming. Sheep and goats provided wool for clothing, milk and cheese, and also meat. Their hides could be used after the animals died. Pigs, easy animals to raise, were a native species to Europe and the Middle East. Pigs also eat most kinds of food scraps and are easy to maintain. The entire animal is edible, and their reproduction rate is extremely high.

Wild cattle were probably tamed and raised or domesticated for the first time in the Middle East. Cattle were adapted to survive in harsh conditions. They provided meat, butter, milk, cream, and cheese as well as hides for clothing. Farmers could harness the energy of cattle to plow fields. Other animals that were used in prehistoric farming

TABLE 1-2	Domestication of Large Animals (Approximate Dates)	
Species	**Date (B.C.)**	**Place**
Dog	10,000	Southwest Asia, China, North America
Sheep	8,000	Southwest Asia
Goat	8,000	Southwest Asia
Pig	8,000	China, Southwest Asia
Cow	6,000	Southwest Asia, India
Horse	4,000	Ukraine
Donkey	4,000	Egypt
Llama/alpaca	3,500	Andes
Bactrian camel	2,500	Central Asia
Arabian camel	2,500	Arabia

included dogs, horses, and donkeys. Donkeys and horses were used for riding and as pack animals for traveling long distances. Dogs were tamed and used to herd and guard domesticated animals and to help in hunting by locating and finding wild animals.

IMPACT ON ENVIRONMENT

In general, the early Stone Age hunter-gatherer societies had neither a large population nor the technical skills or tools to have a major impact on the natural environment. During the Neolithic Period, however, when people turned to agriculture and grazing activities, the local ecosystems began changing. Settlers cut down trees, burned bushes and grasslands, replaced wild plant species with domesticated ones, and diverted water from streams for irrigating crops. As a result of these activities, people started to live in more permanent villages and towns. In time, large populated civilizations were formed from these communities. These early civilizations, discussed in the next chapter, developed their own cultures, languages, technologies, economies, and social values. They also placed more demands on their environment than the hunters and gatherers of the Stone Age.

Vocabulary

Domestication To tame for household use.

Hominids Early human ancestors.

Indigenous Native to a place.

Neolithic Revolution A cultural period beginning about 10,000 B.C. in the Middle East.

Opaque Not letting light through.

Sustainable To keep in existence.

Technology The application of science or the control of the natural world for the benefit of humans. Science is the understanding of the natural world and should not be confused with technology.

Activities for Students

1. Diagram a Stone Age food web centered on humans.

2. Invite a museum employee to demonstrate how tools were made from rocks. Read *Flintknapping: Making and Understanding Stone Tools* by John C. Whittaker.

3. Study the geological locations where flint, chert, quartz, and obsidian are natural resources.

How does this pattern compare to the locations of indigenous peoples and their migration patterns?

4. Create a picture story of a day in the life of a hominid using only naturally found pigments and canvas. Study which natural pigments are the most permanent over time. What effect do light and moisture play on that permanence?

Books and Other Reading Materials

Bahn, Paul G., ed. *The Story of Archaeology*. New York: Weidenfeld and Nicolson, 1996.

Brown, Robin C. *Florida's First People: 12,000 Years of Human History*. Sarasota, Fla.: Pineapple Press, 1994.

Fagan, Brian. *People of the Earth: An Introduction to World Prehistory*. 8th ed. Reading, Mass.: Addison-Wesley, 1996.

Johanson, Donald C., and Blake Edgar. *From Lucy to Language*. New York: Simon and Schuster, 1996.

Kooyman, Brian. *Understanding Stone Tools and Archaeological Sites*. Albuquerque: University of New Mexico Press, 2001.

Whittaker, John C. *Flintknapping: Making and Understanding Stone Tools*. Austin: University of Texas Press, 1994.

Websites

American Museum of Natural History, http://www.amnh.org

Old Stone Age, http://www.oldstoneage.com/

Smithsonian Institution, http://www.si.edu/

CHAPTER 2

Early Human Civilizations

The development of agriculture during the Neolithic Period allowed people to abandon much of their nomadic existence and live in permanent settlements. In these settlements, humans no longer had to live entirely by gathering or hunting for food but to learn how to produce it. These early settlements were established near water sources, fertile soil, and important natural resources.

The inhabitants of these early communities built permanent shelters, cultivated a reliable food source, and created tools for agricultural use. They developed crafts that would enhance their living conditions. Creating a life for people that includes permanent residence, cooperative work efforts, specialization and division of labor, and the establishment of cultural traditions is called sedentism. The ancient civilizations of the Asian and African continents were among the earliest to form these kinds of permanent communities. This chapter discusses some of these early civilizations.

CHALLENGES TO SEDENTISM

These early farming communities faced many social, economic, and environmental issues. They had to overcome the challenges of living closely together. Problems of sanitation, disease, balancing the conflict between wants and needs, sharing resources, and distributing land and wealth had to be resolved. Large numbers of people living closely together also created special problems that called for cooperation among the people of the community. Major sources of concern were the education of the youth, the establishment of rules of order, the division of labor among the workers, and a willingness to protect and defend the community and to develop traditions that would sustain its culture.

Overcoming Environmental Barriers

The first agricultural revolution, which began about 10,000 years ago, brought about a profound change in how human beings perceived the land and its resources. The idea of humans controlling the environment and dominating the *ecosystem* was a major reversal from that of

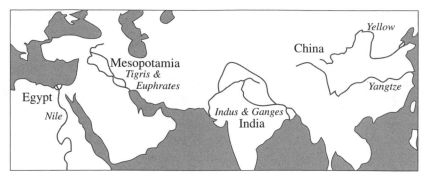

Figure 2-1 • The Four Old-World River Valley Cultures The major ancient communities were usually settled along rivers. The soil of the riverbanks was very fertile for farming and raising livestock. Settlers could fish along the riverbanks and used the waterways to travel by boat, which was a faster mode of travel than by foot.

Refer to Chapter 3 for more information about the second agricultural revolution.

the simpler life of the hunters and gatherers. When human beings began to plant and harvest crops and to domesticate animals for their use, populations increased forming complex societies. Intellectual and technological achievements, literacy, concepts of government and land ownership, and a division of labor and craftsmanship characterized these communities.

MESOPOTAMIA

Mesopotamia, called the Cradle of Civilization, was one of the first agricultural based civilizations. It was established on the *flood plain* of the Tigris and Euphrates rivers in what is now the country of Iraq. In time the people discovered that farther south in the river valley there were natural *levees* that contained silt from the river floods. This silt was good soil for crop cultivation. The marshes along the rivers provided excellent habitats for birds and fish as well as forage for goats and sheep. It was in this region, called the Fertile Crescent, that the early farmers settled, prospered, and eventually formed large cities.

Controlling the Water

As populations increased more demands were made on the environment. Crop yield had to be increased, and the rivers' waters had to be controlled.

The climate of the region allowed two crops to be grown and harvested each year. Farmers were challenged to devise ways of controlling the river water so that it could be used when needed. Irrigation canals and flood control dams were constructed to supply the proper amounts of water when and where it was needed. Clay was used to line the reservoirs of water so that the water would not leach into the soil or evaporate before it could be used to irrigate the fields.

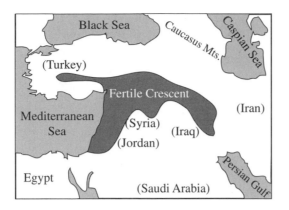

FIGURE 2-2 • The Fertile Crescent encompassed sites of food production before 7000 B.C. It was an area situated between the Tigris and Euphrates rivers of present-day Iraq. Farmers could grow a large variety of crops because of the rich soil and natural resources.

Food Sources

Wheat and barley were the earliest crops cultivated by the farmers of the Fertile Crescent. They added legumes such as chickpeas, lentils, and fava beans to the diet. Over time they cultivated other foods that enriched and balanced their diet: figs, apricots, almonds, eggplant, various nuts, dates, olives, and grapes.

A plant called flax was grown to produce linen fibers, and sesame seeds were pressed to produce cooking oil. When pressed, olives provided oil for cooking and acted as a fuel source for small household lamps. Grapes and barley were used to make alcoholic drinks. Domesticated goats and sheep provided milk, cheeses, meat, wool, and hides for leather.

Development of Metal Tools

At first farmers used stone hoes and wooden plows pulled by oxen. Advanced metalwork in copper and tin produced a stronger, more durable metal called bronze. This period in history is called the Bronze Age. Bronze tools were a major advancement over wood and stone implements. This new alloy was much harder than stone or wood but, most important, it could be shaped into a variety of sizes for tools and weapons. The development of the bronze-tipped plow about 2800 B.C. made turning over hard soil a much less difficult task. Bronze was also used to make spears and other tools.

The first wheeled vehicle is believed to have been used in the Mesopotamian city of Sumer in about 4000 B.C. It was a solid wooden cart with an axle and wheel design. It made hauling material many times easier than a sled or pushcart.

A device called a seeder-plow was invented which not only cut a furrow in the ground but also dropped seed into the soil at the same time. The basic design was not unlike the seeding devices invented in the nineteenth century in England.

One of the most effective and important tools on the farm was the pickax. It was used to break up the sun-baked soil, dig trenches, and prepare the soil for plowing. The grinding stone was an important

Endangered Marshlands of Mesopotamia

The vast marshlands of the Tigris-Euphrates River system once covered nearly 20,000 square kilometers (7,722 square miles). Today less than 2,000 square kilometers (770 square miles) remain. Habitats that existed at the time of the ancient Babylonians are in jeopardy of becoming extinct. Since 1970 extensive damming and irrigation conducted along the rivers has dried up 90 percent of the marshlands downstream. Ecosystems harboring humans, mammals, fish, and waterfowl have been destroyed. What little remains is barren, salt-encrusted desert.

advance in producing large quantities of flour from the hard, dry kernels of wheat and barley. Craftspeople developed techniques for tanning skins. Hides from goats, sheep, and pigs were tanned into shoes, sandals, and water bags.

IRON

The widespread use of iron occurred about 1000 B.C. in the Tigris-Euphrates Valley. This metal was harder than bronze and could be more easily sharpened and shaped into many forms for use on the farm. Iron chisels, hammers, saws, anvils, and drills were common tools found on a typical Mesopotamian farm.

Clay and Bricks

Brick making and basket weaving were two important crafts that used the natural resources available in the environment. Using water and the clay soil found along riverbanks, clay workers produced bricks used for building blocks and pottery for the storage of oils, grain, and water.

Marsh reeds were used in the brick-making process to stabilize the clay while it dried. Reeds were mixed into the moist clay before it was molded into brick-like shapes. The reeds were also gathered and woven into baskets and storage containers. Clay pipes were molded and buried underground to remove liquid waste from populated areas—a first-known attempt at plumbing and waste management. Clay tablets were produced which scribes imprinted with cuneiform symbols to keep records and tell stories.

Economy and Trade

In addition to the cities' requirement to meet basic human needs—food, shelter, and clothing—their desire for additional goods and services meant that they had to extend themselves beyond the confines of their own communities.

One way to meet these needs and desires was to engage in trade. Trade enabled cities to exchange goods and natural resources with other cities. It also gave people an opportunity to meet in the marketplace where they could share ideas and information.

The Euphrates River was the main artery for trade. It was called Uruttu, meaning "copper," because copper and other minerals such as iron and lead had been discovered in the Zagros mountain region to the north and east. The river was used to transport these raw materials. These natural resources, as well as grain, spices, cotton cloth, wool, and leather, were traded in the marketplaces of the cities located along the river. Many cities, such as Babylon, became very wealthy as a result of this river trade.

The Legacy of the Fertile Crescent

The land of the Fertile Crescent was made productive by people who observed and analyzed their geography, identified and exploited their natural resources, and developed the means to control their land and water sources. This allowed for the development of farming techniques which fed its population and provided a surplus that ensured that a growing population was possible.

People began to specialize in craftsmanship, construction, government and law, soil and plant science, and the arts. For the first time, a writing system was developed to permanently record the knowledge acquired by the people of these early cities. Exchanges with other cultures led to new ideas in the sciences and technology and cultural advancements in the arts and literature, laying the foundation for other civilizations to build upon.

Human Impact on Mesopotamia's Natural Resources

The successes of the largest city-states in Mesopotamia—Sumer and Babylon—reached their peak about 2500 B.C. At about that time, the large system of canals and ditches, which had been used to carry water from the Tigris and Euphrates rivers to the fields, was beginning to become contaminated by salt, a condition called *salinization*. Ocean water was backing up into the irrigation system and depositing small amounts of sea salt on the land. When the water evaporated the salt remained on the fields. After 5,000 years of growing barley and wheat, the fields became too salty to grow crops and had to be abandoned.

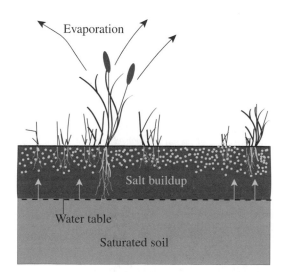

FIGURE 2-3 • Salinization of soil occurs when soluble salts such as sodium chloride, calcium chloride, and magnesium chloride accumulate in soils to a level that harms plants and prevents their growth. Salinization caused the Mesopotamia farmers to abandon their fields after many years of good farming. Salinization is still an environmental problem in many farming communities today.

The coastal areas along the Mediterranean were heavily forested. The timber was used for smelting iron and for building. Many forests were cut down without planting new trees. This process is known as *deforestation*. Today few trees grow in modern Lebanon and Israel.

The competition between the city-states for control of trade, land, natural resources, and freshwater often led to bloody conflicts. Eventually, these wars weakened the cities, drained them of people and raw materials, stopped the exchange of goods, and finally resulted in their mutual demise. By 500 B.C. the Persian army had conquered the last of the Mesopotamian cities—Babylon and Ninevah—and the era of the powerful city-states of the Fertile Crescent was at an end.

Refer to Volume IV for more information about deforestation.

ANCIENT EGYPT

Geography and Natural Resources

From its earliest beginnings, in about 3000 B.C., life in Egypt has depended greatly on its greatest natural resource: the Nile River. The Nile, 6,650 kilometers (4,160 miles) long, is the longest river in the world. In ancient times Egypt's population settled along the flat river basin and the delta it formed as it emptied into the Mediterranean Sea.

Soil Resources

The Nile Valley can be separated into two very different parts: what the ancient Egyptians called "The Black Land" and "The Red Land." The black land was the fertile strip of silt along the banks of the Nile River which resulted from the river's annual flooding. The red land was the vast, dry desert area that bordered both sides of the fertile riverbanks. The rich, dark earth deposited each year by the Nile's floodwaters created a band of fertile soil, which provided early farmers with nutrient-enriched soils in which they could cultivate their crops.

Water Resources and Irrigation

For farming communities to survive along the river, the Egyptians had to devise a system that would control the floodwater of the river and, at the same time, provide water for crops during the dry season. They learned how to build irrigation ditches and canals. They built dams upstream to control the flow of the river and calculated ways to predict the quantity of floodwaters that would inundate the fields.

Sluice gates were built to control the water going directly into the fields. These gates raised and lowered the water levels in the canals and irrigation ditches. A device, called a "shaduf," was invented to lift water in buckets over the banks of the canals to move water into the fields.

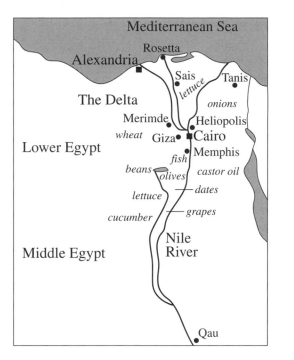

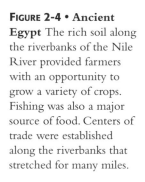

FIGURE 2-4 • Ancient Egypt The rich soil along the riverbanks of the Nile River provided farmers with an opportunity to grow a variety of crops. Fishing was also a major source of food. Centers of trade were established along the riverbanks that stretched for many miles.

Waterwheels and a device patterned after Archimedes' screw was also used to raise water out of the river.

Agriculture

As the Egyptians became more able to control their environment, they were able to grow crops twice a year. This greatly increased their agricultural production.

There were three seasons to the agricultural year: flood time, seedtime, and harvest time. During seedtime, which occurred between November and February, the cropland was laid out and marked, canals were constructed to hold the water deposited by flooding, and the fields were plowed and seeded. The plow was pulled by oxen, and the seed was broadcast over the soil by hand. Farm animals were then let loose on the moist, seeded soil so that their hooves would drive the seed into the soil.

At harvest time, from March to June, everyone worked the fields. They used sickles to harvest the wheat. As soon as the grain was bundled together, it was taken to the threshing floor where animals trampled on the grain until the grain pods fell off the stalks. Then it was winnowed: tossed into the air where the lighter chaff was blown away leaving the grain. It was then stored for future use.

Other crops harvested during this time were beans, lettuce, peas, radishes, onions, olives, dates, grapes, and cucumbers. Two kinds of oil, castor and sesame, were also produced for cooking and lighting. Birds, fish, eggs, and some beef and pork supplemented their diet.

Refer to Volumes II and IV for more information about agriculture.

Other Natural Resources

The Nile River provided not only the rich soil for the farmers but also raw materials for many other purposes. The papyrus reed found along the banks of the river provided material for baskets, ropes, sandals, boat construction, mats, and a crude form of paper. Another weed, flax, was used to make a very lightweight, durable cloth material called linen. The Nile River also supplied the Egyptians with natural sources of food; wild fowl and fish were very plentiful.

Rock quarrying was an important task performed by the Egyptians. It occurred during the time of flooding when little work could be done in the fields. Limestone was quarried for use on the outside of buildings. Workers used copper tools to saw and chisel the stone. The stones were moved by using human power to pull them onto barges, which floated them on the river to where they were needed. Sleds were then used on watered ground to slide the stones to where they would be placed.

Another resource, river clay, was mixed with weeds, formed, and then sun-dried to create building blocks for smaller structures.

Economy and Trade

The cities along the Nile became centers for trade throughout the region. Boats called feluccas with their adjustable sails were able to maneuver up and down the gently flowing river with goods for trade. Overland caravans coming from the centers of eastern civilizations carried a variety of goods: iron, precious metals, ivory, rare woods, animal hides, incense, and spices. All of these materials were traded in the marketplaces of the cities along the river and the coastline of the eastern Mediterranean Sea.

Gifts and Benefits of the Early Egyptians

Sedentism in the lives of the ancient Egyptians resulted in the spectacular construction of pyramids, temples, and statues. The Egyptians produced a written language (hieroglyphics), a yearly calendar, mechanical engineering, astronomy, and advanced mathematics. Their technological and artistic skills formed the basis for other civilizations to modify and improve upon. They created a civilization that contributed a great many gifts to the modern world.

Human Impact on Egypt's Environment

Water was an important part of the sanitation process in Ancient Egypt. Sanitation systems were set up to dispose of household garbage and waste. The irrigation canals were the main disposal areas for dumping waste. However, the canals became a breeding ground for disease and a habitat for vermin. Water pollution was a major threat to the health of the human population of the Egyptian cities. Air pollution in the form of soot and smoke occurred mainly in the large cities. Evidence of smoke-darkened marble on some buildings has been recorded.

ANCIENT GREECE

Geography

Although no part of Greece lies very far from the sea, most of the landscape is mountainous with deep, isolated valleys. Mainland Greece is a peninsula reaching out into the Mediterranean and the Aegean seas. Many islands surround the mainland.

Unlike the settlers along the rivers in Mesopotamia and the Nile in Egypt, the early Greeks and their Phoenician neighbors were unable to establish large agricultural communities. Both of these civilizations lacked the necessary natural resources to support their populations. Only about 20 percent of mainland Greece is suitable for farming, and most of that is located along the coast. The lack of *arable* land and rainfall, which occurs mostly in the winter months, made farming very difficult. Ancient Greeks found it necessary to develop skills in navigation, boat building, and engineering to exploit the resources located on other islands in the Aegean Sea.

Early Farming

The land and the climate restricted agricultural activities in much of Greece; however, small family farms were established along the coast and near small river systems. Farming is believed to have begun in Greece about 6500 B.C. with wheat and barley among the first of the crops to be cultivated. Eventually, grapes, lentils, and beans were added to the agricultural landscape. Cultivating olives was an especially important job because olives provided food, cooking oil, lamp oil, and soap.

Animals were bred and raised for milk products, meat, and hides. Pigs, goats, sheep, poultry, and some cattle and oxen could be found on a typical small Greek farm. In order to use the land more efficiently, the farmers used a technique called *terracing*. In this land-use technique, steps are cut into the land on the hillsides and each step, or terrace, is cultivated.

Irrigating crops called for devices that could lift water in buckets above the level of the river or water source and spill the water onto the fields. Waterwheels, designed much like a modern Ferris wheel, were built to accomplish this task.

Farming in ancient Greece was a very labor-intensive endeavor. Much of the farm family's work was assisted by slave labor. Slaves played a significant role in the everyday operation of the farm. It has been estimated that one-third of the population in Athens and the nearby farms were slaves.

Terracing is an agricultural method of reshaping the hilly terrain in order to grow crops and prevent soil erosion. Terracing was practiced in several ancient civilizations and continues to be used today in many countries. (Courtesy of Lynn Betts, USDA, NRCS)

The Lost City of the Incas, called Machu Picchu, is near Cuzco, Peru. The Incas became self-sufficient in food production by terracing the slopes and filling them with transported soil. (Courtesy of Jane Mongillo)

City-States

In about 750 B.C. Greek villages began to join together to form larger communities called a polis or city-state. This was an independent city with its own rules of government, culture, and traditions.

By 450 B.C. the city-state of Athens had grown wealthy through its trading operations in the eastern Mediterranean. It had established itself as the major trading center in the region. Very much like a modern shopping mall, the agora, or marketplace, was the busiest part of the city. More than just the buying and selling of goods took place in the agora. It was a place for people from many different parts of the

TABLE 2-1	Aegean and Eastern Mediterranean Natural Resources	
Gold		Copper
Iron		Silver
Fish		Timber
Grain		Lead
Marble		Honey
Horses		Oil

known world to meet and talk, share ideas, and practice their religious beliefs at the temple.

Environment and Natural Resources

The natural resources available in the eastern Mediterranean area from the Black Sea to north Africa and Italy were important to the seafaring Greeks and Phoenicians. Iron, copper, gold, silver, and lead were mined throughout the region. Timber was harvested on the shores of Crete, Macedonia, and Lebanon. Marble was quarried on the southeast coast of mainland Greece. Honey, fish, salt, and hemp were found in Thrace and Sicily.

Economy and Trade

The colonies and outposts that were established provided the city-states of Greece with the trade goods and raw materials needed and wanted by the population. Wheat, olives, figs, wood, linen, dried fruits, salted fish, lumber, hides, precious metals, honey, wine, and pottery were among the myriad number of goods bought and sold in the marketplaces.

The Legacy of the Greek Civilization

The Greek civilization left behind the concept of a democratic government, the idea of a large centralized marketplace offering many kinds of activities for its customers, mechanical devices for construction purposes such as levers and pulleys and other simple machines, a universally accepted written and spoken language, and an appreciation for creative work in the arts and literature.

Human Impact on Greece's Environment

The intense competition for natural resources and finished products from around the eastern Mediterranean Sea resulted in environmental changes, piracy, and conflicts between and among the city-states and foreign empires.

Mining and quarrying operations scarred the land. Greek hillsides, once covered with trees, were nearly barren because of the enormous amounts of wood needed for smelting iron and other ores. There was a loss of vegetation on many farms caused by farmers clearing the land for grazing purposes. Overgrazing caused the loss of plants and in turn resulted in severe soil erosion on what was once rich, fertile land.

The dilemma of societies attempting to satisfy unlimited wants with limited resources often led to confrontations between them. In 338 B.C. Greece was badly weakened by internal fighting among its city-states and continued warfare with the Persian Empire. It became unable to defend itself and was defeated by the armies of Macedonia and fell under the control of Alexander the Great.

ANCIENT ROME

The ancient city of Rome was located on the Italian peninsula that stretches into the Mediterranean Sea. Rome's location on the Tiber River and at the center of the Italian peninsula and its closeness to the Mediterranean Sea offered major advantages to the Romans. The location could easily be defended, the soil was fertile, and the river water was available for agriculture. The city's easy access to the Mediterranean Sea was another advantage.

The height of the power of the Roman Empire was established about A.D. 100. The large empire eventually stretched from the British Isles, the Iberian peninsula, and North Africa, across southern Europe, to the countries of Egypt and the eastern Mediterranean.

Agriculture and the Roman Empire

Throughout the Roman Empire, farming was the main occupation of most of its people. Many farmers were located in the rich, fertile valley of the Po River, located in the north. Farm products were used for personal consumption and also as a means of trade and paying taxes to the landowners and Roman rulers.

AGRICULTURAL TECHNOLOGY

As the need to feed an expanding population grew, new ideas evolved to increase farm production. Iron was used to improve the efficiency of plows, sickles, and hoes. Irrigation techniques were engineered so that canals and drainage systems were made more dependable for watering crops.

Roman farmers learned from the Greeks about the importance of fertilizing the soil. Techniques were developed to use animal and vegetable waste on the fields. The manure was applied to the fields by means of a canal system using water to spread it over the soil. The farmers also used mineral fertilizers, made from saltpeter and potassium nitrate, to enrich the soil.

AGRICULTURAL PRODUCTS

Wheat, barley, olives, and grapes were the main crops common to the Italian peninsula. As the empire expanded foods from other lands were imported to Italy to be grown in its fertile valleys. These agricultural imports included new kinds of wheat, apricots, lemons, melons, sesame, flax, various nuts, and fruit trees.

Natural Resources

Metal ores found in the countryside near Rome were iron, copper, zinc, and tin. These represented nearly all the valuable minerals located in Italy and the central part of the Mediterranean area. A large variety of natural resources originated throughout the Roman Empire, however. Some examples include the following:

- British Isles: tin and iron

- Gallia (France): gold

- Hispania (Spain and Portugal): gold and copper

- North Africa: timber, marble, herbs, and papyrus

- Eastern Mediterranean: frankincense, iron, glass, fish, horses, ivory, pepper, asphalt, and precious stones.

DID YOU KNOW?

Thousands of years ago, salt was traded as a resource with as high a value as gold. The ancient Roman armies called one of their major highways the Via Salaria: the Salt Road. The Roman soldiers were paid a salarium, or salary. This was money given to them for exceptional work.

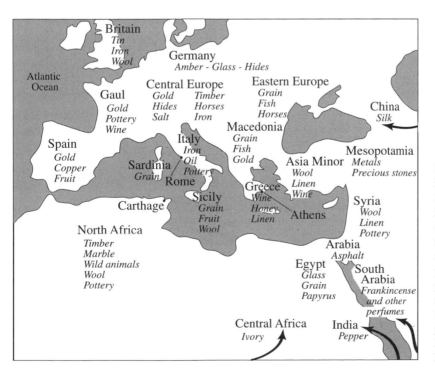

FIGURE 2-5 • Economy in the Roman Empire During the Roman Empire much of its economy consisted of farming, fishing, and the trading of goods. Every part of the Roman Empire provided a variety of manufactured goods for trade that included glass, precious stones, animal hides, pottery, jewelry, linen, ivory, gold, and tin.

Economy and Trade

Trade was extensive throughout the Mediterranean. Sea routes and roads built by Roman engineers and protected by the Roman military crisscrossed the entire region. Goods of all kinds came into the marketplaces from every corner of the empire. Wool, linens, perfumes, grain, precious stones, pottery, honey, hemp, wild animals, wine, fruits, salt, and animal hides were among the commodities bought and sold throughout the empire.

Technology

Roman engineers built and maintained an enormous network of roads, aqueducts, and bridges. The road system consisted of more than 80,000 kilometers (50,000 miles) of roads throughout the empire from Asia Minor to the British Isles. They carried trade goods as well as the Roman armies.

Aqueducts were built to bring freshwater to the inhabitants of the cities, many of which had polluted the river systems and needed *potable* water. Their discovery of how to make concrete allowed them to build structures that were much taller than had been possible before. Many of their roads, bridges, and aqueducts are still in use today, some 2,000 years after their construction.

Benefits of the Romans

The ancient Romans were extraordinary builders of roads, bridges, aqueducts, and public buildings. They established a system of laws based on what was "good and fair" and created the Latin language which forms the root for many modern European languages. Much of what they achieved in technology, the arts, and literature still influence us today.

Human Impact on the Roman Empire

As the empire grew so did the city. It is estimated that the population of Rome at its peak approached a million people. As a result of this population growth, it had many of the same problems faced by most large cities today. It was overcrowded, dirty, and noisy, and it suffered from major traffic congestion. Air pollution, resulting from local smelting businesses, was also a problem. For the poor, it was not a pleasant place to live.

Over time, the city's water supply, the Tiber River, was so contaminated that potable water had to be brought into the city from long distances by way of large aqueducts designed and built by Roman engineers and slaves. The city had a sewer system to carry away waste, but it was not adequate to carry away the large amounts of waste produced by the animals and human inhabitants of the city. Garbage was often thrown into the streets; human waste and household trash were often dumped out of the windows of three- and four-storied wooden apartment-type buildings. Fire, always a danger, was an ordinary event in these crowded slum areas.

The Roman government did provide open spaces for the enjoyment of its citizens. Public parks and baths were established where people could relax, play ball, wrestle, and enjoy the grassy fields. Stadiums were built to provide entertainment for the people, but it had become clear that the once proud city was beginning to deteriorate.

As the population of Rome increased, more demands were made for grain and vegetable products to feed the people. This required an extensive importing of foods, especially wheat, to feed the inhabitants of the city.

The increasing need for food put pressure on the Roman leaders and the army to expand their influence and power over overseas territories so that the necessary agricultural resources could be provided for its citizens. This put a strain on the army and its ability to protect the empire. It was becoming too large to control and was being challenged by barbaric tribes on its frontiers. The city of Rome was overrun by Germanic tribes about A.D. 400.

Besides environmental issues, the decline of the Roman Empire and its influence in the Mediterranean world can be attributed to several factors:

- Corrupt and weak government leadership

- Increasing numbers of poor people in the population depending on a growing scarcity of food

- Fewer sources of new wealth and available natural resources in the region

- Decline in the value of the monetary system

- Military pressure from northern tribes.

ANCIENT CHINA

Geography and Climate

The geography of China is made up of four distinct regions: the high plateau and the Himalayas mountain range to the west and southwest; remote desert areas in the north, northwest, and northeast; the low north China plain; and the warm southeast coast. China's geography has been a source of its isolation for many centuries. As a result of this isolation, the people of China developed a language, customs, and technology separate from the influence of the civilizations that had been established in Mesopotamia and the Nile River Valley.

The enormous expanse and variety of the Chinese landscape leads to a great diversity in its climate. The cold winters of the northern and western mountains, deserts, and plateaus contrast sharply with the warm, moist climate of the southeast. Northeastern China has a temperate climate; summers can be warm, but winters are cold and snowy.

SOIL AND WATER RESOURCES

China's major river systems, the Huang Ho (Yellow River) and the Yangtze (Chang Jiang), represent the two dominant areas where ancient Chinese people established permanent agricultural settlements in about 5000 B.C. The fertile soil of the north China plain, called loess, was unique because it was soft and porous and easily cultivated with simple tools.

AGRICULTURAL PRODUCTION

The earliest crops grown were cereal grains called millet. Soybeans were cultivated and became a source of protein for the Chinese diet. It was also used as a fertilizer to enrich the soil in preparation for growing millet and, eventually, wheat. They grew a variety of vegetables, such as Chinese cabbage, radishes, eggplant, peas, yams, sugar cane, bamboo shoots, and leeks. Where these crops were grown successfully depended upon the climatic conditions of the area.

Rice was grown in the warm, moist climate of the Chang Jiang River basin. The early Chinese developed methods of flooding their fields to form rice paddies. As the rice grows, its roots must be covered by water. Artificial dikes and deliberately planned lowland river flooding accomplished the task. The farmers worked small plots of land and therefore did not keep many large animals. There was not enough arable land to allow for much grazing.

The Chinese farmers, over the centuries, have always strained to feed a growing population. With limited arable land available for agricultural production, they have constantly looked for ways in which to improve crop yields.

The farmers invented iron-tipped plows, wheelbarrows, and a collar and harness mechanism to control animals while working the fields. They devised a system of irrigation that used pedal power and buckets to raise water out of the river to water the fields. Growing silkworms, or sericulture, was a major agricultural enterprise. Silk, an important source of income, was traded as far west as the Mediterranean Sea.

FIGURE 2-6 • Map of the Silk Road The Silk Road consisted of several trading routes from the Middle East to China. This map shows two of the major routes that Mediterranean traders used to travel the 7,000 kilometer-long Silk Road journey to China.

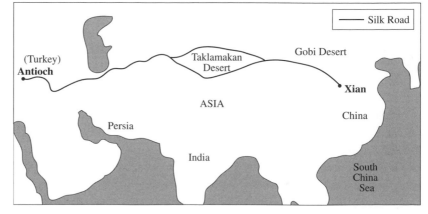

Economy and Trade

Ancient China could not provide for all the needs and wants of its people. Although large in territory it had little arable land, which was located in the eastern third of the empire. The need for trade became increasingly important to China's leaders.

Many dynasties ruled China over the course of many centuries. The Han Dynasty, which came into power in 202 B.C. had, perhaps, the most dramatic effect on ancient China. Under the leadership of the emperors in this dynasty, China's borders were extended westward through the deserts and mountains to central Asia. Trade routes were established with communities along the routes, and goods such as silk, hides, fruits, ivory, fabrics of wool and linen, jade, and items of bronze and precious metals were traded.

Eventually, several of these routes joined forming the Silk Road, which was about 7,000 kilometers (4,200 miles) long. The major trade routes that made up the Silk Road extended from the Chinese city of Xian in the east, along the Great Wall in the north, bypassing the Gobi and Taklamakan deserts where possible, west across the Pamir Mountains, through Mesopotamia, and finally ending in the eastern Mediterranean at the city of Antioch in Turkey. From there trade goods were bought and sold throughout the Mediterranean region.

Technology

The people throughout China were forced to work on large construction projects such as roads and canals. The largest, most enduring structure built was the Great Wall across northern China, which is estimated to be about 2,400 kilometers (1,400 miles) long. Its main purpose was to keep out the marauding Mongols from the north.

Another major project was the building of the Grand Canal, which connected the two major Chinese Rivers: the Hwang Ho and the Chang Jiang. This canal is still one of the largest waterways ever built. It is about 1,600 kilometers (1,000 miles) long and is called the "transport river" by the Chinese. The canal was used as a transportation and trade route between the two most productive areas of ancient China.

Legacy of Early China

The early Chinese people contributed much to the world's knowledge. They developed iron farm implements, the wheelbarrow, silk production, moveable type, printing paper, a land compass, ship rudders, and coal for heating. They also made inroads into the study of the human body systems.

Human Impact on China's Natural Environment

Much of China's population has been concentrated in the river valleys and coastal plains. As China's population grew more land was needed for agricultural use. In order to acquire additional land for crops widespread cutting of trees took place. Over time, this resulted in severe deforestation and the desertification of the grasslands along the river floodplains.

The lack of trees and grasses to anchor the soil caused silt to run off into the rivers. This not only resulted in soil erosion but also caused a buildup of silt which raised the water level of the river. Often the rivers would overflow their banks and cause severe flood damage with serious loss of life and property.

Dikes and dams have been erected along the rivers for centuries; however, they have not always been able to control the river water. Thousands of lives have been lost to floods. The Huang Ho River, the one most prone to flooding and the destruction of life and property, has been called China's "river of sorrows."

ANCIENT INDIA

Geography and Climate

The geography of India caused its people to be isolated from much contact with the civilizations that had been established in Mesopotamia, China, and the Nile River Valley. The high Himalayas, Hindu Kush mountains, the Arabian Sea, and the Indian Ocean all separated the early settlers of the Indus River Valley from much of the known world.

The climate of India is controlled by strong winds that blow across the subcontinent. These winds bring dry air to the region during the fall and winter months. By late spring the winds reverse direction and move across the Indian Ocean picking up moisture. These winds, called monsoons, bring great amounts of rain onto the land.

Too much rain can cause flooding which devastates people, property, and crops. Not enough rain where it is needed can cause a

FIGURE 2-7 • India and the Monsoons
A monsoon is a major wind system that changes its direction during a seasonal change. In India there are two distinct monsoon seasons. The summer monsoons bring heavy rain for the important growing season; however, too much rain can cause massive flooding conditions. Winter monsoons are usually cold and dry, and the dry harsh winds can cause droughts and damage to crops.

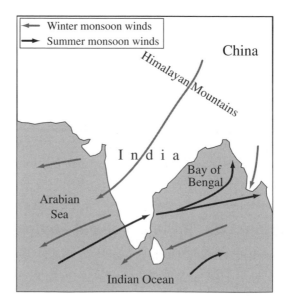

drought which reduces the production of the farmland and endangers the food supply.

Early Legacy

Not much is known about the civilizations that were established along the Indus River. The pictographic writing left behind by the Harappans has remained a mystery to archeologists. Scholars find it difficult to decipher the early writings of these people, making knowledge of the Harappa culture very difficult to obtain.

What is known is that they established cities with streets, sewers, baths, and pools. They worked with bronze and copper to make pots, pans, farm implements, and weapons. They knew how to form gold, silver, and jade into decorative pieces and are even thought to have traded with the city-states of Mesopotamia.

Early Agriculture

Mohenjo-Daro and Harappa were two of the first cities established in the Indus River Valley. The settlers were an agricultural people whose life and economy were almost completely taken up with food production. They reached their highest population growth between 2500 and 1500 B.C.

At first the crops grown in the valley were rice, cotton, and sugar cane. When techniques were devised to dam and drain the river water for irrigation, other crops were introduced. Wheat, barley, sesame, and dates became part of the diet. Pigs, goats, sheep, and cattle were among the first animals to be domesticated. About 1000 B.C. the horse, donkey, and Bactrian camel were domesticated and put to use for herding purposes and beasts of burden.

Sickles made of specially worked stone were used for harvesting the crops. Iron was eventually discovered. It was smelted and made into tools for crop cultivation and for weaponry. The wooden plow was used extensively to prepare the soil for planting. The largest building in the community was the granary, a place for storing grain. Farmers took their grain there as a protection against famine and as a payment of taxes.

Human Impact on India's Natural Resources

The settlements along the Indus River plain began to decline in about 1500 B.C. People began to leave the cities and abandoned their fields. Archeologists who have studied this region believe that soil exhaustion caused by overcropping, the expansion of the neighboring desert, changes in the course of the river, flooding of the fields, and, finally, invasion by outsiders were the major factors causing people to leave.

Vocabulary

Arable Land useful for cultivation.

Deforestation Removal of trees without replanting to make arable land.

Ecosystem System which includes all organisms of an area and the environment in which they live.

Floodplain Plain or flat landscape bordering a river that can be subject to flooding.

Levee Embankment built to hold back water.

Potable Water that is safe to drink.

Salinization Process by which soil becomes more salty.

Terracing Cutting hill slopes to form terraced fields to prevent soil erosion and to monitor and control irrigation.

Activities for Students

1. Create a topography map for each season mapping the effect that the seasons had on the Tigris and Euphrates as well as on the Nile River Valley.

2. Go to a local museum and information center to found out about the first settlement of your current town or city. What geographical features contributed to the success of the settlement?

3. Create a chart demonstrating and comparing the evolution of the tools used by farmers in their sedentary new lifestyle in Mesopotamia and the Nile River Valley from clay to bronze, 4000 to 1000 B.C.

4. Using the Three Little Pigs as a model, build three structures out of different organic materials (clay, mud, straw, stones, sand, etc.) to discover the best ingredients to create a permanent, lasting structure. What effects do moisture and light play on their permanence?

5. What geographical advantages did Rome and Greece have that Mesopotamia and the Nile River Valley did not which led the former two countries to develop as centers of trade?

Books and Other Reading Materials

Danforth, Kenneth, ed. *Journey into China*. Washington, D.C.: National Geographic Society, 1976.

Fabor, Harold. *The Discoverers of America*. New York: Charles Scribners and Son, 1992.

Martell, Hazel. *The Vikings*. New York: Warwick Press, 1986.

Milton, Joyce. *Sunrise of Powers, Ancient Egypt, Alexander and the World of Hellenism*. Boston: Boston Publishing Company, 1986.

Pennington, Piers. *The Great Explorers*. London: Bloomsbury Books, 1979.

Severy, Merle. *Greece and Rome: Builders of Our World*. Washington, D.C.: National Geographic Society, 1968.

Speake, Merle. *Atlas of the Greek World*. New York: Facts on File Publications, 1982.

Van Loon, Hendrik. *The Story of Mankind*. New York: Liveright Publishing, 1967.

Websites

Exploring World Cultures, http://eawc.evansville.edu

42 Explore, www.penncharter.com/student

Just Curious: Ancient Civilizations, www.suffolk.lib.ny.us/youth/jcancient.html

Metropolitan Museum of Art, www.metmuseum.org

The search engine Google will access the preceding sites. You can also search with the key words "ancient civilizations" or "middle ages" for additional sites that may be of value.

CHAPTER 3

The Agricultural Revolution: Expansion of Agricultural Productivity

Very little changed in agricultural techniques and farm production from the time of the ancient civilizations, such as the Roman Empire, to the Middle Ages. By the 1700s, however, significant events had occurred that necessitated more productive farming methods. A large growth in population signaled the need for increased food and textile production. Specific strategies were required to

- Manage the production of farmlands more efficiently

- Design and build better farm tools and machinery

- Experiment in *selective breeding* of plants and animals

- Find newer and safer methods of food preservation

- Establish reliable transportation systems to distribute farm products.

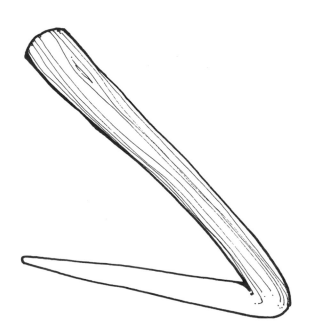

FIGURE 3-1 • The early digging tools were made of bone or wood and were used in spading the soil to grow crops.

EXPANSION OF AGRICULTURE

The population increase in much of Europe during the Middle Ages (A.D. 900–1400) called for a major expansion of agricultural growth. Several factors combined to create the setting for an increase in agricultural productivity. The climate had undergone changes that produced milder winters, more consistent rainfall, and a longer growing season. Land considered useless for cultivation, such as swamps, were drained. Forests were cleared, and lowlands near the sea were recovered and put under cultivation.

When the horse began to be used as a major source of power, farmers invented the horseshoe, harness, and collar to improve their control of the animals. Wind and waterpower were used to move waterwheels and windmills for grinding grain. A wider use of iron in farm implements increased the efficiency of the plows and hoes. Early efforts in animal breeding and more efficient irrigation techniques added to the productivity of the farm.

England was in the forefront of these events. Many of the new developments in farming methods and in tool and machinery manufacturing took place beginning in the late 1600s and eventually spread to America by the nineteenth century.

ENGLAND AND THE SECOND AGRICULTURAL REVOLUTION

Early Agricultural Production

Refer to Chapter 2 for more information about the First Agricultural Revolution.

In about A.D. 1000 the English manorial system of agricultural production and land distribution was put in place. This system allowed *serfs*, under the protection of the lord of the manor, to farm small strips of land in commonly held fields. The serfs were also able to graze their livestock with those of other serfs in community grazing areas.

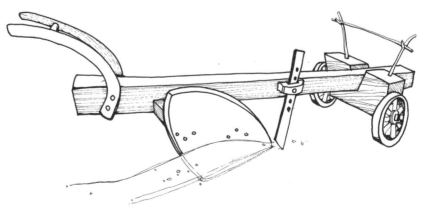

FIGURE 3-2 • The mouldboard plow, used in the Middle Ages, was a major farming tool that could plow large tracts of land for crops. This plow could also be used on land that was quite wet. Besides human energy, horses or oxen were used to pull the plow.

By the fifteenth century the owners of these manors or estates recognized that this system of agricultural production was inefficient and unprofitable. Ineffective crop rotation caused nutrients to be depleted from the soil. The quantity and quality of crop production declined. Inbreeding of commonly held livestock did not produce stronger or healthier sheep or cattle. Not enough land was available for experimentation that might lead to improved crop production and strengthened livestock breeding.

Time for a Change: Land Enclosure

In order for the owners to make the land more profitable, they eliminated the manorial system and began "enclosing" the land under single ownership. Peasants and small farmers were dismissed, hedges were used to enclose the land, and more pasturage was created for sheep production. This system of land enclosure meant that larger estates could be maintained, crop production increased, and pastureland extended to provide for additional wool production and, consequently, increase profitability for the landowner.

Crop Rotation and Natural Fertilizers

The traditional system of leaving a field *fallow* for a year, thereby losing the production of that field, was replaced by a four-course rotation of crops. This change, developed by Sir Charles Townsend, put nitrogen and other nutrients back into the soil. This system of crop rotation was on a four-year cycle. In alternating seasons the land was planted with wheat, barley and legumes, and clover and turnips for cattle feed. This eliminated the practice of having part of the arable land inactive since

Crop rotation reduces the need for the extensive use of fertilizers and pesticides by alternating crops. (Courtesy of Robert Kolberg, USDA-ARS)

A History of Fertilizers and Pesticides

Since the beginning of the agricultural revolution thousands of years ago, farmers have been aware that two factors greatly affect the growth and quality of crop production. They knew that sufficient soil nutrients were necessary for healthy plant growth and that plant-eating insects could destroy a crop in a very short period of time. As a result of their knowledge, they experimented with fertilization and pest control.

Fertilizers

Wood ash was one of the first natural fertilizers used to enrich the soil. The use of human and animal manure was also found to be effective. By the seventeenth and eighteenth centuries, crop rotation using legumes and grasses became part of an agricultural cycle that improved the quality of the soil. This technique put nitrogen elements back into the soil as plant nutrients. Other natural fertilizers such as guano and a mixture of lime, clay, and gypsum called marl were used. Each of these natural fertilizers was an effective means of enriching the soil.

In the late nineteenth and early twentieth centuries, scientists discovered that other chemicals were needed by plants to improve their growth: phosphorous and potassium. Farmers soon added these chemical fertilizers to their fields as well as the traditional natural ones.

Eventually, using chemical fertilizers containing all of the plant nutrients crops needed became the dominant approach to replacing soil nutrients. Over time, however, scientists discovered that the excessive use of these chemical fertilizers was a recognizable source of environmental pollution.

Pesticides

Throughout recorded history, insect pests have affected the comfort, health, and agricultural production of human beings. From the earliest days of the farmer, a variety of methods have been used to protect crops from insect infestation. Some of their methods worked, and others had questionable results which often impacted the environment more than the insects and pests.

In 300 B.C., the Greeks used "brimstone" or sulfur as a fumigant to get rid of pests. Other societies used a myriad of other techniques. Combinations of pepper and tobacco, soapy water and vinegar, and turpentine and lye were used with marginal success.

The early Chinese used highly poisonous arsenic to control garden pests. Fish oil was used as a fungicide. In the mid nineteenth century, Paris Green—a deadly mixture of arsenic and mercuric chloride—was used to destroy soil-inhabiting insects. Another arsenic-based poison was used to control a beetle that was destroying potato crops. Hydrogen cyanide, whale oil, and kerosene were all used to spray fruits and citrus trees.

In the early twentieth century, copper sulfate and calcium cyanide were sprayed on wheat and cotton crops. In 1939 the chemical DDT was introduced as an insecticide, and its use quickly became widespread. In later years it was found to be extremely toxic to humans and wildlife. The use of this insecticide has been banned, but its effect on wildlife and the food chain was devastating.

all sections of the fields were producing a crop for a particular purpose—either animal feed or food for human consumption. Because this four-course system provided food for livestock, meat-producing animals were available all year.

Natural fertilizers such as *guano*, lime, and a combination of limestone and clay called marl were used extensively. Manure from animals grazing in the turnip and clover fields was an additional source of fertilization.

Refer to Volume IV, for more information about the Second Agricultural Revolution.

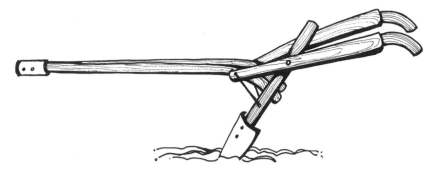

FIGURE 3-3 • The shovel plow did not have wheels in the front like the mouldboard plow. The innovation of various designs of the plow allowed the farmer to harvest not only food for his or her family and livestock, but produced a surplus to be sold in the marketplace.

Farm Mechanization

The improvements seen in agricultural production in England in the eighteenth century were due directly to the advances made in industrial mechanization in the factories. The development of the time- and labor-saving devices used in the factories and mills to increase production were adaptable for use on the farm. Farmers, scientists, and inventors began to experiment with the use of steam power and the application of various mechanical concepts to improve the efficiency of farm equipment. There were several early successes in adapting mechanized techniques for use on the farm. The first farm implements using industrial processes were powered by horse or oxen.

The horse-drawn seed drill replaced the inefficient and wasteful practice of broadcasting seed by hand. Funnels in the drill drove the seed below the surface of the soil ensuring that seed was buried. Fewer seeds were likely to be blown away or eaten by animals or insects. Sturdier farm implements, such as plows fashioned from iron rather than wood, could be kept sharper and were more durable. The development of the mechanical reaper replaced the hand-swung sickle and scythe, allowing for the harvesting of larger quantities of grain in a shorter period.

Ideas borrowed from industrial applications influenced agricultural technology. Water-, wind-, and steam-powered machinery could grind grain to flour, saw wood, channel water, plow fields, and harvest crops. The farmer who was able to use the new technology increased farm production and made agriculture a profitable enterprise.

Livestock Breeding

Experimentation in domestic livestock breeding became a major function on many of the farms. Livestock breeding, especially with sheep, was first pioneered by Robert Bakewell in the late 1700s. His breeding techniques permitted more wool production per sheep. He was also successful in breeding sheep for food as well as for wool production. Other breeding techniques were developed to increase animal size, strength, milk production, and taste.

AGRICULTURAL REVOLUTION IN AMERICA

Slash-and-Burn Farming

The slash-and-burn technique, the main form of land clearing for crop production used by the indigenous people of southern New England, caused major changes in the ecosystem of forested areas. First, the tree bark was girdled or slashed. Then dry wood was piled against the trees and set on fire. This process of slashing and burning eventually kills the trees. Over a period of from five to seven years, the trees die and fall making it easier for them to be moved. The ashes from the burnt trees and surrounding vegetation provided nutrients for the soil.

Indian corn, called maize, as well as squash, beans, and pumpkins were grown around the remaining tree stumps. In a few years, when the soil became depleted of nutrients, the land was abandoned and left to grow other types of natural vegetation. This process of land clearing and subsistence farming was repeated throughout the forested areas.

Refer to Volume II for more information about indigenous people.

The result was a forest of large, widely spaced trees which allowed more sunlight to reach the forest floor. The additional sunlight enabled the growth of food such as berries, which attracted game animals. The native people hunted these animals as an addition to their food supply.

Early Colonial Farmers

Few of the colonial settlers in America's forests used slash-and-burn techniques for clearing their land. They used the axe and plow to cut the trees and plow the soil. This work was very labor intensive and took up much of the farmer's day.

Settlers allowed their livestock to wander about the forest eating what they could find, much as wild game would do. Cattle and hogs thrived on this technique and provided an easily maintained and more reliable meat supply than hunting. A farm was considered successful if several hectares (acres) of corn could be grown and enough farm animals produced to supply food to eat with a small surplus for trade.

AGRICULTURAL EXPANSION IN AMERICA

The early farmers of New England and the middle Atlantic area cultivated small farms of about 8 to 10 hectares (20 to 30 acres). At first, farm technology in early colonial America consisted of extensive hand labor and the use of rudimentary tools. Axes, hoes, and plows were typical early farming implements. Seed was broadcast by hand, and grain crops were harvested by sickle and threshed with a flail. Not much more than one-half hectare (acre) could be harvested in a day.

After the American Revolution, improved farming techniques and tools borrowed from the advances made in agricultural production in Europe allowed farmers to increase their cultivated land. A typical farm of 40 hectares (100 acres) might include 12 hectares (30 acres) of

Slash and burn is a forestry practice that involves cutting down large tracts of trees to clear the land for farming and ranching activities. (Courtesy of Rainforest Action Network)

cropland, 8 hectares (20 acres) of pasture, and the rest left to woodland and brush. Increased farm production often resulted in surpluses, which were used in trade throughout the region.

AGRICULTURAL PRODUCTS OF THE NEW NATION

Agriculture in New England consisted of dairy farming and mixed vegetable, corn, and fruit production for home use and for sale in local markets in *urban* areas close to the farms. Wheat flourished in the farms of the mid-Atlantic States. The cultivation of cotton and tobacco became cash crops for the Piedmont and Southern states; much of it was sold in overseas markets.

Other important crops and livestock consumed at home or grown for commercial trade along the East Coast, the Southern states, and the islands of the West Indies were potatoes, sheep primarily for wool, flax for linen cloth and linsey-woolsey (a cloth combination of linen and wool fibers), hemp for rope and sailcloth, pigs and cattle for lard and meat products, and sugar cane for rum and molasses.

The Great Plains: A New Region to Farm

HOMESTEAD ACT OF 1862

In 1862 President Abraham Lincoln signed into law the Homestead Act. This act gave American citizens the opportunity to own 65 hectares (160 acres) of land in the Great Plains for a $10 filing fee. If the land was farmed for a period of five years, the farmer received the land free. Most of the original homesteads were located along river drainage systems where there was an adequate supply of freshwater and some trees. As a result of the Homestead Act, large numbers of farmers and others moved to the Great Plains.

The plains of North America can be divided into two regions. The dividing line is considered to be the 100th longitude or meridian line. Areas east and west of this imaginary line have distinct differences in precipitation, natural vegetation, and soil varieties.

EAST OF THE 100TH LONGITUDE The average rainfall for this area of the Great Plains is from 8 to 10 centimeters (20 to 25 inches) per year. Warm air currents moving northward from the moist atmosphere of

FIGURE 3-4 • The Great Plains and Prairies

the Gulf of Mexico account for much of the precipitation on the east side of the 100th meridian. This movement of moist air creates a climate pattern favorable for crop production.

For thousands of years the natural vegetation and grasses played a significant role in creating rich, organic soil on this part of the plains. Tall prairie grasses reaching a height of about one meter (three feet) dominated the landscape. As the grasses grew and decomposed they returned valuable nutrients to the soil. This created layers of grass, soil, and roots which formed a thick mixture called "sod." Although the sod was difficult to plow, it proved to be an ideal soil for growing a variety of crops. Between 1865 and 1900, the number of hectares in wheat, corn, oats, and cotton doubled or tripled in size. The number of farms in this region during the same time period increased from 2.6 million to 5.3 million farms.

WEST OF THE 100TH LONGITUDE West of the line, referred to by early explorers in the area as the "Great American Desert," The Great Plains was a desolate, treeless, dry area of grassland that appeared to offer little promise for agricultural success. Any farming done on this side of the longitudinal line needs considerable irrigation to be successful. However, the short grasses that grow in the region are ideal for cattle ranching.

AQUIFERS AND AGRICULTURE

The Ogallala *aquifer* is the major water source for much of the agriculture in the Great Plains. It is the largest freshwater underground water source in the world, covering approximately 450,000 square kilometers (174,000 square miles) underneath the land surface. Irrigation using the Ogallala aquifer began about the mid–1870s, at about the same time that steam-driven water pumps came into use.

By the 1920s manufacturers had developed hydraulic irrigation systems which eventually led to sprinkler and drip irrigation. With special piping and controls, both of these methods used less water, and the water flow from the pipes could be controlled more than the flood and furrow irrigation. These two types of irrigation, which controlled the movement of water pipes over the landscape, deposited water on the plant roots. This was especially useful on the rolling plains of the Midwestern states.

Farm Technology

Early attempts to use the agricultural technology developed in England in the eighteenth century were not generally successful in the American Colonies. Rocky soil or previously uncultivated fields made the seed

Refer to Volume IV, for more information about the Ogallala aquifer.

The Original Occupants: The Native Americans

The original occupants of the Great Plains were the Native Americans and the American bison, or the buffalo. Attracted by the native grasses growing on the plains, the buffalo numbered in the millions. Many tribes including the Kiowa, Pawnee, Cheyenne, Crow, and Sioux followed the herds. Hunting the buffalo was the primary activity of these tribes. These huge animals provided them with food, clothing, and tools: nearly everything they needed to survive.

At first, the tribes had to follow the buffalo on foot, making this source of food inconsistent for weeks at a time. When the Spanish explorers left the area in the sixteenth century, they left behind horses which the American Indians learned to domesticate. They used the horses as a rapid means of hunting the buffalo. Also, horses allowed the tribes to follow the migration of the animals ensuring a more consistent source of food.

DID YOU KNOW?

On the plains, violent thunderstorms, tornadoes, blizzards, hailstorms, drought, grasshoppers, and locusts could destroy a farmer's work at any time during the year. Many farmers were not able to endure the hardships of the plains and abandoned their farms. Others stayed and persevered. These farmers established on the American Plains the "breadbasket" of the world.

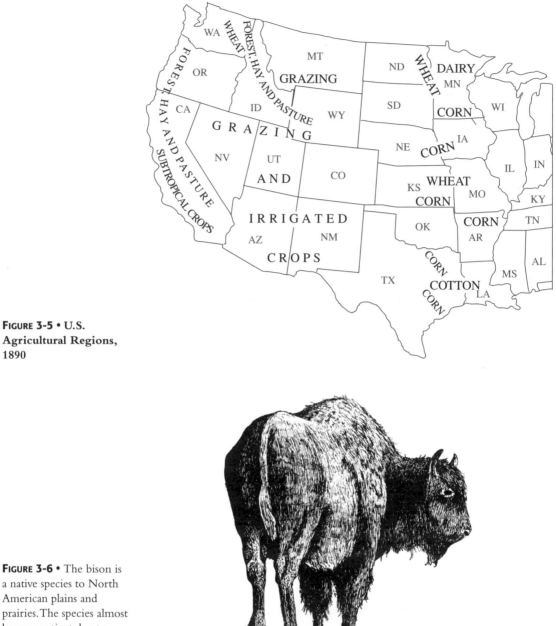

FIGURE 3-5 • U.S. Agricultural Regions, 1890

FIGURE 3-6 • The bison is a native species to North American plains and prairies. The species almost became extinct due to over hunting in the 1800s

drill and the horse-drawn mechanical reaper impractical. By the 1820s, the cast-iron plow was first put to use on the prairie sod of the Midwest. The John Deere plow with its self-sharpening steel edge was an important improvement in turning the thick soil of that region.

A horse-drawn planter was developed in the 1850s. This device dropped seed automatically in evenly spaced rows much like Jethro Tull's seed drill. Until 1860 the scythe was still the basic tool used for reaping the harvest. Cyrus McCormick was working on a mechanical

The John Deere Prairie Queen plow was introduced about 1874. John Deere helped revolutionized farm equipment in the mid-nineteenth century. Deere built a stronger, lighter and more efficient steel plow for prairie grasses. (Courtesy of Deere & Company Archives)

Eli Whitney's cotton gin played a major role in the establishment of the textile industry in the United States. His machine replaced the labor-intensive process of separating seeds embedded in the cotton by hand. The cotton gin separated the cotton seeds much faster allowing more production of cotton fiber in less time. The cotton gin in the photo was used on a Louisiana cotton plantation. (Photo by Peter Mongillo, taken by permission of Louisiana State Museum, Cabildo Museum, New Orleans)

reaper, and by the end of that decade 70 percent of the wheat grown in the plains states was harvested by this machine. A two-horse treadmill, devised by Hiram Pitts, could be brought to the field for threshing and binding the grain.

The threshing machine in the photo is harvesting wheat near Spokane, Washington, in the early 1900s. The threshing machine, as well as other improvements in farming equipment, allowed farmers to produce crops for commercial markets around the world. (Courtesy of the Library of Congress)

The invention of the steam engine was used in producing new farm equipment, such as tractors. The steam tractor, such as this 1917 Case Steam Tractor, replaced animal-powered energy machines on the farm. (Courtesy of Heritage Photographs)

Horsepower was the main source of energy used to run these machines. A harvesting combine invented by Hiram Moore and powered by from 12 to 16 horses could cut and thresh up to 10 hectares (25 acres) of wheat per day. However, using horsepower of this size was not always practical. The disadvantage of using horses was their inconsistency in powering the machines in terms of speed and control. Often fields were too small to accommodate the number of horses and the machinery needed to work at this pace.

The invention of steam and, later, internal combustion engines eventually made sowing and harvesting machines practical for most farms.

Improving Traditional Crops

For many centuries farmers looked for ways to improve the quality and productivity of their crops. At first environmental conditions and soil fertility determined the yield and quality of agricultural products.

By the mid–1800s the work of botanist Gregor Mendel began to influence agricultural production. His work on plant *genetics* recognized that plants inherited specific characteristics. This knowledge influenced how farmers approached the improvement of their crops. Ideas of plant breeding led to the development of crops that were more resistant to pests, able to tolerate harsher climates, more resistant to diseases, and were larger and tastier. The farmers on the Great Plains adapted their crops to the climate and prairie soil of the region. Turkey Red wheat and hybrid types of corn were planted and produced healthier, more productive yields per acre.

Scientist George Washington Carver and his research institute in Alabama made major breakthroughs in the production of peanut

Asian farmers have been growing soybeans for 5,000 years. Soybeans contain about 40 percent protein and are used for human consumption, edible oil, and feedstock for animals. Soybeans are used in more than 300 products. (Courtesy of Agricultural Research Service, USDA)

products, sweet potatoes, and soybeans—all important products serving a variety of uses. Other scientists have developed other high-yield varieties of wheat, corn, barley, and sorghum appropriate to various climates and soils.

BENEFITS OF THE AGRICULTURAL REVOLUTION

New methods of agricultural production were being discovered during the 1700s, at about the same time another revolution was taking place: the Industrial Revolution. Technological advances in mining and textile manufacturing called for large numbers of workers increasing the population in urban areas. More people needed to be fed.

With the aid of scientists and inventors, farmers developed ways to increase their food quality and production and ways in which they could transport their agricultural products more quickly and more efficiently. These advances in agricultural production were being made particularly in Europe and North America.

Applying new ideas from the natural sciences to *horticulture* and animal breeding helped develop better, more diversified crops and increased crop yield per hectare (acre). Animals were bred for specific purposes. Larger animals were bred for transportation, farmwork, quality of meat, and leather goods.

Farm machinery evolved from the use of the hand-held sickle and scythe to horse-drawn and then steam- and gasoline-powered engines which drove seed- drillers, plows, reapers, and threshers. These improvements in agricultural production, coupled with improved methods of transportation and safer techniques for canning, bottling, and refrigeration, made possible the movement of perishable food products over long distances.

A tractor is spraying pesticides on a cotton crop. (Courtesy of United States Department of Agriculture, photo by Bill Tarpenning)

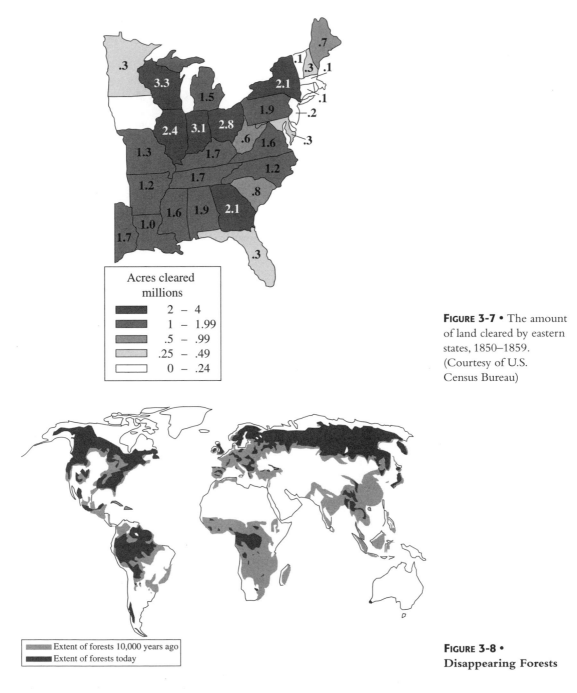

Acres cleared
millions

2 – 4
1 – 1.99
.5 – .99
.25 – .49
0 – .24

FIGURE 3-7 • The amount of land cleared by eastern states, 1850–1859. (Courtesy of U.S. Census Bureau)

Extent of forests 10,000 years ago
Extent of forests today

FIGURE 3-8 •
Disappearing Forests

Not all of the farm products were needed for domestic consumption. As technological improvements were made in transportation and food preservation surplus agricultural products were exchanged with other countries. This exchange allowed for the increase in the amount and diversity of the foods consumed by people throughout the world.

By the end of the nineteenth century, the United States had become the breadbasket of the world. Currently, the United States is the leading exporter of wheat and corn.

Dust Bowl

The United States experienced its greatest loss of topsoil from erosion in the late 1930s and early 1940s, during an event that has come to be called the *Dust Bowl*. The Dust Bowl disaster resulted from severe wind erosion in the southern Great Plains region of the United States caused by a combination of poor agricultural practices and years of sustained drought.

Beginning in 1931, powerful dust storms carrying millions of tons of black soil repeatedly swept across western Kansas, the panhandles of Texas and Oklahoma, and the eastern portions of Colorado and New Mexico, rendering millions of hectares (acres) of farmland useless. The devastation resulting from this disaster awakened people to the need for understanding the factors that can lead to increased erosion. Soil erosion practices are now followed to prevent such losses in the United States and in many other countries.

FIGURE 3-9 • Dust Bowl Region

AGRICULTURAL IMPACT ON THE ENVIRONMENT

Temperate North America suffered much deforestation during the Second Agricultural Revolution. Trees were cut down to provide additional land for cultivation or livestock grazing. Cut-and-burn techniques and overgrazing can alter forests so that the land becomes useless for growing crops and grazing animals. When the forest is first cleared, the nutrients in the soil are sufficient to support crop growth and grazing. Within three or four years, the nutrients are used up and the grass and crop yields become smaller, often competing with invading weeds for survival.

When the land can no longer produce crops, the deforested land is abandoned and another part of the forest is cut and the process repeats itself. The abandoned land is left with sparse vegetation resulting in vulnerability to wind and soil erosion and the continued loss of plant and animal habitats. The entire ecology of an area can be affected. Other negative impacts on the natural environment included the excess runoff of fertilizers and pesticides into water resources and the draining of wetlands.

Refer to Volume IV, for more information on soil erosion and the Dust Bowl period of the United States.

Hugh Bennett was an American soil scientist and advocate for soil conservation methods in agriculture. Bennett gave many speeches to farmers during the 1920s and 1930s on the importance of soil conservation. (Courtesy of National Archives and Records Administration)

Vocabulary

Aquifer Porous rock or soil which water flows through and in which water gathers to supply wells.

Dust Bowl A large area of farmland in a number of states (Kansas, Oklahoma, and Texas) which lost topsoil through wind erosion and drought in the 1930s and early 1940s.

Fallow Soil that has been plowed but not seeded during the growing season.

Genetics The study of the way in which characteristics of an organism are inherited through genes.

Guano Bird droppings used for fertilizers.

Horticulture Growing of plants for food or decoration.

Selective breeding Selection of individual plants and animals with useful characteristics for breeding future generations.

Serfs Members of the poorest and lowest class of society during the Middle Ages of Europe (A.D. 1000–1400).

Urban Densely populated city area.

Activities for Students

1. Build a waterwheel and a windmill to generate enough power to light a lightbulb.

2. Visit local farms and interview farmers about the ways in which they replenish the nutrients in the soil and how they breed their livestock.

3. In the expansion of the west, finding water was essential. Research the divining rod which some people used as a tool to find water underground. How practical and useful is the divining rod?

4. Read about the Donner party and their wagon train adventure. What supplies did they take with them for their journey?

5. Build a greenhouse and grow various plants in it. As fungi and pests develop, experiment with different organic pesticides such as brimstone, sulfur, pepper, tobacco, and soapy water.

Books and Other Reading Materials

Ambrose, Stephen, and Douglas Brinkley, eds. *Witness to America*. New York: HarperCollins, 1999.

Birdsall, Stephen, and John Florin. *Outline of American Geography, Regional Landscapes of the United States*. New York: John Wiley and Sons, 1992.

Burchell, S. C. *Age of Progress*. New York: Time Incorporated, 1966.

Cronon, William. *Changes in the Land*. New York: Hill and Wang, 1991.

Diamond, Jared. *Guns, Germs and Steel*. New York: Norton, 1999.

Ferris, Robert G., ed. *Prospector, Cowhand and Sodbuster*. Washington, D.C.: National Park Service, 1967.

Mitchel, Robert D, and Paul A. Groves, eds. *North America: The Historical Geography of a Changing Continent*. Boston: Roman and Littlefield, 1990.

Schlereth, Thomas J. *Victorian America*. New York: HarperCollins, 1991.

Sutherland, Daniel. *The Expansion of Everyday Life, 1860–1876*. New York: Harper and Row, 1989.

Wolf, Stephanie Graumen. *As Various as Their Lands*. Fayetteville: University of Arkansas Press, 2000.

Websites

Exploring World Culture, http://evansville.net

Geography Home Page, http://geography.Miningco.com/library

History Link 101, www.historylink101.com

United States Department of Agriculture, www.usda.gov

CHAPTER 4

The Industrial Revolution: Economic, Social, and Environmental Changes

For thousands of years, the Neolithic Revolution had produced a global society and a subsistence economy tied to agriculture; however, by the 1700s new technologies and methods of producing goods gave impetus to another revolution, the Industrial Revolution. This new era started in the late 1700s in England and the early 1800s in the United States and Western Europe.

The Industrial Revolution was marked by an increase in the exploitation of Earth's natural resources, the development of new sources of energy, faster and more efficient manufacturing techniques, and new means of transporting goods, people, and information. These events resulted in dramatic changes in how and where people lived, how they earned a livelihood, how they satisfied their wants and needs, and how they interacted with and altered their environment.

PRE–INDUSTRIAL REVOLUTION

Prior to the Industrial Revolution, fewer than 10 percent of the people of Europe lived in cities. Most of the population was located in rural areas, and nearly all of their productivity and resources came from agricultural jobs. Energy came primarily from wind, water, and human and animal muscle power. Transportation on land was by horse and on water by sail. Communication was very slow and depended upon post riders, hand-printed newspapers, and word of mouth from travelers.

The small towns and villages spread throughout the countryside were self-contained, subsistence societies. People spent most of their time working their farms and raising crops for their own use. What little surplus they had was traded in the local marketplace. They used the natural resources they found in their environment to make their own furniture, farm tools, cooperage, metal goods, and clothing.

Cottage Industry System

In towns and cities specialized craftspeople manufactured various products. Organizations of craftspeople, called guilds, established rules to control the quality of workmanship, the number of items that could be produced and sold, the prices of items produced, and procedures as to who would be allowed to become members of the guild.

These craft workers, who used specialized but simple hand tools, made items of silverware, cloth, leather, hardware, jewelry, and weapons. Since the guilds controlled the quality, quantity, and price of the goods, their products could be very expensive.

Merchants, on the other hand, wanted to sell items more cheaply and in larger quantities than the guilds would allow. They looked for other methods of manufacturing and found it in the hands of the rural farmers and their families.

The merchants distributed raw materials to the farmer's home or cottage, and the farmer and his family manufactured the finished product. Along with the responsibilities of maintaining successful farms, farmers produced clothing, textiles, wood products, metalwork, and food products. They were paid for their work, and the merchants took the finished products to town in hopes of finding a market for these goods.

INDUSTRIAL REVOLUTION IN ENGLAND

The Factory System

By the eighteenth century, the population in the towns and cities had increased dramatically as births outnumbered deaths. The need to manufacture more goods to service this growing urban population became the impetus for devising more productive, more efficient machines and machine tools. Some *entrepreneurs*, merchants, and business owners thought that organizing workers together under one roof would be more efficient and profitable than the cottage system. The new method of organizing workers in a manufacturing enterprise was called the factory system. The merchants no longer had to travel to distribute raw materials to individual craft workers or cottages and then travel back to pick up the manufactured goods. The new system allowed merchants to use *mass production* strategies which produced more goods on a much larger scale and at a cheaper price. The factory system ultimately increased the profits of the owners.

Luddites

Not everyone embraced the changes that were occurring in textile production. In 1812 the Luddites, a group of English weavers, rebelled against the use of power looms in the mills. They argued that the machines produced an inferior product and that they had caused the elimination of their jobs. They formed into bands of men who broke into mills and destroyed the new machines. After several incidents of bloodshed, the government cracked down on the Luddites hanging some and imprisoning others.

The Birth of the Industrial Revolution

The Industrial Revolution, which began in Great Britain, had spread to most of Western Europe and the eastern United States by the middle of the nineteenth century. Great Britain's lead in industrialization was a result of several advantages it had over other countries: geography and climate, sources of raw materials, global influences, financing, a reliable labor force, and reforms in land ownership.

TABLE 4-1	Great Britain's Economy	
Resources	**Imports**	**Exports**
Coal	Rubber	Beef
Iron ore	Ivory	Mutton
Salt	Metals	Sugar
Gypsum	Hemp	Tea
Fertilizer	Cotton	Wheat
Granite	Tobacco	Spices
Lead	Coconut oil	Finished cotton goods
Clay	Cocoa	Leather goods

Beginning in the mid–1700s in Great Britain, the Industrial Revolution altered how people lived, worked, and purchased goods and services. Great Britain was the world leader in the Industrial Revolution with a very strong economy in imports and exports.

Great Britain's location as a large island north of mainland Europe had several other advantages. Surrounded by the sea and with an irregular coastline, no British industry needed to be far from the seaports. Industries could be established along inland waterways, and finished products could be moved to accessible markets and ports with little difficulty.

Great Britain has a marine west coast climate. The mild wet climate is responsible for the growth and abundance of a wide variety of crops and is very suitable for the successful raising of livestock, especially sheep and cattle.

Refer to Volume I for more information about climate conditions and ecosystems.

Resources of Great Britain

Great Britain had rich resources of coal, iron ore, salt, agate, gypsum for plaster and fertilizer, granite for building, lead for piping, clay, and water. It also had colonies overseas, which supplied Britain with rubber, ivory, metals for coinage, hemp, jute and sisal for rope making, palm and coconut oil for paints and lubricants, cotton, and tobacco. Food was also imported from Australia, the East Indies, China, South America, and the Caribbean. Great Britain exported mutton and beef, sugar, spices, wheat, and tea to help feed their growing population. The country also possessed a strong navy and a merchant marine fleet which were able to sail to ports all over the world trading finished goods for raw materials.

Great Britain as a World Leader

The Industrial Revolution in Great Britain took manufacturing out of the cottages and workshops. Machinery replaced handwork and handmade tools. Factories were established to bring together under one roof the worker, the machines, and the raw materials. The success of the factory system that replaced cottage industries caused Great Britain to be recognized as a powerful force on the world stage as a leader in coal mining, iron and steel production, woolen and cotton manufacturing, and shipbuilding.

INDUSTRIAL REVOLUTION IN AMERICA

Samuel Slater—Father of the American Industrial Revolution

Samuel Slater has been called the Father of the American Industrial Revolution. In 1790 he successfully designed and built the first textile factory on the Pawtucket River in Rhode Island. This water-powered mill spun, carded, and drew out cotton thread in preparation for producing textiles. Soon Slater's idea about using water to power textile mills spread to other industries in New England. The American textile industry began to compete with Great Britain for world markets. Slater had established the factory system in America and inspired inventors and other entrepreneurs to invest in that system.

Other industries using water power soon followed the factory system models established in the Lowell mills in Massachusetts and the Slater mills in Rhode Island, laborers gathered in one workplace to operate power-driven machines that manufactured finished products from raw materials.

Edmund Cartwright and Others

In 1785 Edmund Cartwright patented a power loom which applied mechanical power to hand looms. At first the power loom used water-power to run the machinery; later, steam power was utilized for the

This steel plant in Ensley, Alabama, in the early 1900s and many other U.S. steel plants during that time accounted for much of all steel castings and ingots manufactured in the world. Steel production was an important part of the Industrial Revolution economy. (Courtesy of the Library of Congress)

Samuel Slater is considered to be the founder of the cotton textile industry in the United States. The Slater textile mill in Pawtucket, Rhode Island, used waterpower to run the cotton spinning and weaving machines. (Courtesy of Slater Mill Historic Site)

same purpose. The power loom could weave material much faster than a hand loom. By 1813 there were 2,400 looms in operation; by the 1860s, 500,000 looms were working.

Inventor Eli Whitney and Samuel Colt were enhancing mass-production strategies. The two men changed the way in which work was performed in gun factories. They developed machines that produced the same part more accurately than hand-made fittings and coupled that idea with an assembly-line operation. In this system, the worker needed to perform only one or two tasks as the gun moved to final assembly. This technique of mass production was eventually put into use in most factory systems.

The invention of the sewing machine by Elias Howe (1846) meant that garments could be sewn more efficiently and quickly than garments made by hand at home or in tailor shops. The production speed of the sewing machine resulted in more finished garments being sold in the markets.

The American Factory System

By the mid-1800s, the manufacturing of textiles was the high point of industrialization in the United States. New England became the center of textile manufacturing. Although not rich in natural resources, the region had innumerable quick running streams and rivers to power the mills. The humid climate kept the cotton and wool fibers manageable for easier spinning and weaving of the cloth.

The most famous textile mills, located in Lowell, Massachusetts, employed the most mill workers, many of whom were young women from the countryside. The mills were among the largest, cleanest, and most technically advanced operations in the world.

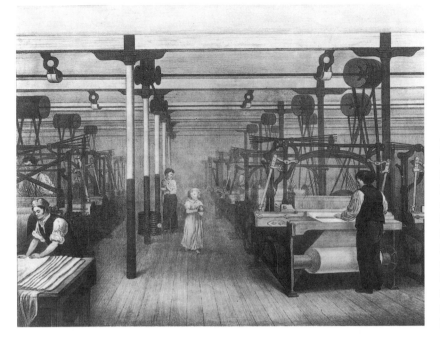

FIGURE 4-1 • By the mid-eighteen hundreds, the Slater Mills made the New England region the center of the cotton textile industry in the United States. More efficient factory-based weaving machines replaced early home-based hand spinning and loom machines. (Courtesy of Slater Mill Historic Site)

Mill Girls

In 1836 20 major textile mills were located on the Merrimac River in Lowell, Massachusetts. These mills employed about 7,000 workers, most of them females between the ages of 15 and 30. The girls were recruited by representatives of the mill owners from rural areas of Maine, Massachusetts, and New Hampshire. These "mill girls" were the first generation of women in America to be employed and earn wages outside the farm.

They were trained as operatives to work at the water-powered machinery in the mills. As operatives they carded, spun, and wove cotton and woolen cloth. They were usually at the mill at 5 o'clock in the morning and worked a 12- to 14-hour day, 6 days a week. The women's lives were regulated by company rules. The representatives who recruited them promised their families that they would be safeguarded. They lived in supervised boarding-houses, were held to a nightly curfew, and were required to attend church services. Any violation was justification for firing the girl and sending her home.

The girls earned about $14 a month less a $5 fee for room and board. The working conditions were difficult, but on their day off the mill girls were able to take advantage of the cultural opportunities available in the city. They attended the lyceum where they heard concerts, lectures, and literary discussions. These activities were not available in the farm communities where they had lived. Most of them used part of their wages to send some money to their families, made purchases for themselves, and saved money for the time when they would quit the mill. The female operatives normally spent only a few years working at the mills before marrying or returning home.

As a result of poor economic conditions in the 1850s, mill owners decreased wages, increased working hours, and demanded increased production. Most of the female operatives left the mills rather than work under these worsening conditions. Immigrants who needed jobs were willing to work and were hired. By the late nineteenth century, the relatively pleasant conditions that had made Lowell an attractive place for women to earn a wage and improve themselves had all but disappeared.

Cotton from the Southern states was sent north to textile mills in Lowell and other mills in New England. The cotton was spun into thread, woven into cloth, and sold to markets throughout North America. By 1860 the New England mills were producing two-thirds of all cotton textiles in the United States.

CHANGE IN ENERGY SOURCES DURING THE INDUSTRIAL REVOLUTION

Muscle energy, both human and animal, had been the main source of power for performing work for thousands of years. As most of Western Europe and North America began to expand in population, they faced the problem of producing more food and more textiles to support it. Different kinds of tools and equipment powered by more efficient and durable energy sources needed to be developed in order to fill this need.

By the 1850s about 50 percent of the energy needed for agricultural and manufacturing purposes was still being met by wood, with coal and waterpower making up the rest. At the beginning of the twentieth century, coal had displaced wood as the primary energy source, with small amounts of oil and natural gas being used in urban areas. By the mid-century, oil and natural gas became the major fuel sources used by the modern industrial countries of the world.

Early Wind Power and Waterpower

The two main sources of power at the beginning of the Industrial Age were wind and moving water. Both of these sources were free and when harnessed to windmills or waterwheels were much more powerful and efficient power sources than muscle power. However, both of these had significant disadvantages. Inconsistent wind patterns and wind speed could result in no movement of the windmill; too much wind damaged the machinery causing breakdowns in mill production.

Although waterwheels were used extensively as a power source for the growing textile and iron goods industries, they too had disadvantages. In order for the machinery in the mills to run, they needed a continuous flow of water to move the large wheels. If water levels were too low or if the wheels could not turn because of frozen water, all production stopped. As the demand for factory goods increased, manufacturers looked for more consistent, more reliable sources of power to run their mills and factories.

Fuel Wood, Charcoal, and Coal

The fuel used for domestic purposes—heating and cooking—was wood; however, the primary use of timber in manufacturing was for the production of charcoal for use in smelting ores, particularly iron ore.

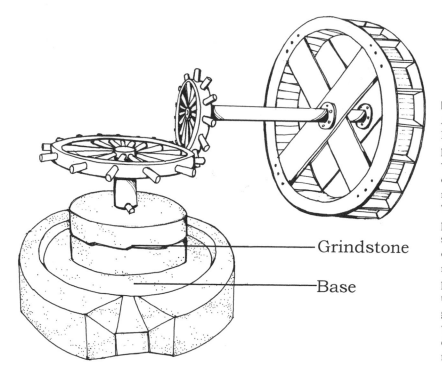

Grindstone

Base

Figure 4-2 • During the Agricultural and Industrial Revolutions, the turning power of a water wheel was used to grind grains on the farm and to run machines in the factories. The waterwheel could be placed in or over the source of a moving stream or river. Even a small, narrow stream could provide enough water to turn the wheel. Special gears attached to the wheels were used to slow down or speed up the flowing current.

For a short period of time, the forests of eastern North American withstood the onslaught of timber cutting, but by the middle of the nineteenth century much of the region had become treeless. Using up timber faster than it could be replaced forced business owners to look for other sources of fuel to feed their furnaces. The coal deposits in the Northeast and the mid-Atlantic areas became the solution to their voracious need for fuel.

Refer to Volume II for more information about fuel wood.

Great Britain and the United States had large resources of coal deposits. The British turned to coal to provide the heat source for steam boilers, furnaces, and smelting activities. In the United States large quantities of bituminous coal were found in Pennsylvania, West Virginia, Kentucky, and Tennessee. Large deposits of anthracite coal were found in eastern Pennsylvania. By the mid-1800s, coal mining and coke manufacturing were important endeavors in these states; most of the coal and coke produced was used for smelting iron ore and manufacturing iron products for the farm and factory.

Iron and Steel

The discovery of coal in Great Britain and in the United States contributed to their world leadership in the industrial production of iron and steel goods. Steel was known to be a material superior to cast or wrought iron.

The process of producing cheap steel was developed by Henry Bessemer in 1850s in England. In the *Bessemer process*, pig iron was

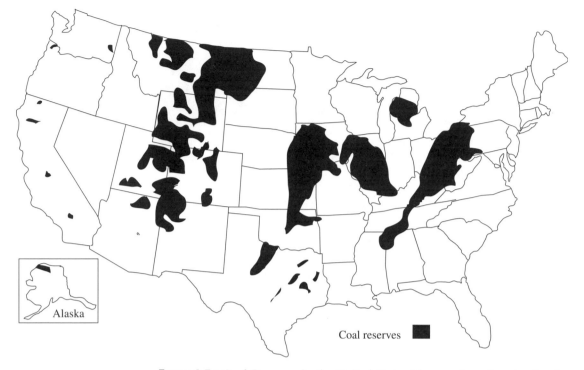

FIGURE 4-3 • Coal Reserves in the United States There are large deposits of coal in the United States. Coal is an essential natural resource for making many products such as iron and steel. The United States has nearly 275 billion tons of recoverable coal. That's more than 250 years of supply at today's usage rates.

Breaker Boys

A coal breaker is a large, tall building located close to the coal mines. Coal cars were pulled by cable from the mine to the top of the breaker. The car was tilted and the coal, along with debris such as slate and rock, was sent down the long chute to the bottom of the breaker. The job of the breaker boys was to separate the coal from the unwanted material, called "culm." The clean coal continued down the chute and into waiting railcars; the culm was dumped onto large piles outside the breaker. Thousands of tons of culm were separated by the breaker boys each year.

The work of the breaker boy was backbreaking, dangerous, and unhealthy. The boys, some of them as young as six or seven, sat on backless benches, and as the coal and debris came down the chute they slowed it down using their feet. Barehanded, they then picked out the unwanted slate and rock. They were not allowed to use gloves because it was thought that the boys' fingers would not be able to move quickly enough to remove all of the debris. This process resulted in swollen and raw fingertips, a condition called "red tips." Even while enduring this pain, they were expected to work efficiently in removing the culm.

The boys worked in terrible conditions. The deafening noise, smoke, and coal dust in the breaker were endured every workday by the boys. They earned from three to six cents per hour and worked a 12-hour day. When they were 14 they were expected to begin work inside the mine. By that time many of them were already showing signs of black lung disease.

Child labor laws passed at the beginning of the twentieth century signaled the end of using small boys in the breaker. It has been estimated that 50,000 breaker boys were employed in the coalfields of Pennsylvania between 1830 and 1910.

melted at extremely high temperatures producing steel, a lighter, stronger, and more flexible metal. Owing to its increased strength and flexibility, steel could be manufactured into railroad tracks, beams for high-rise construction, bridges, steel hulls for oceangoing ships, and military armaments.

Refer to Volume II for more information about coal.

Coal Gas

Coal gas, a byproduct of coke production, was found to be an effective source for lighting purposes. It produced a brighter light than candles or whale oil lamps at a lesser cost.

Gasworks built in Great Britain provided gas lighting for the towns, mills, and factories until the early decades of the twentieth century. Gas lighting provided several benefits: city streets were safer, factories could run longer hours, and homes and schools could be provided with a brighter lighting system.

The production of coal gas also had an important byproduct: coal tar. This product could be used to make a variety of products including artificial fertilizers and artificial dyes. The production of coal gas had a significant impact on the early development of the modern chemical industry.

Natural Gas

The history of using natural gas dates back to the early societies of Greece, Persia, China, and India. These ancient cultures used natural gas as part of their religious services. The Chinese were the first to recognize natural gas as an energy source. They constructed bamboo piping to transport the gas from shallow wells to sites where it was used to heat seawater hot enough to evaporate it and collect the much-needed salt residue.

The first commercial drilling for natural gas occurred in New York State in 1821. Natural gas soon became popular as a fuel for lighting street lamps. It was not used as a fuel source for homes at that time because there were no pipelines to bring the gas to the homes. Some construction of natural gas pipelines was completed at the end of the nineteenth century, but it was not until the 1950s that major gas pipelines were built. Thousands of kilometers of pipelines crisscross America today.

Refer to Volume II for more information about natural gas.

Petroleum Resources

Petroleum oil has been used for a variety of purposes since the time of the ancient Egyptians. The Egyptians used it to seal parts of the great pyramids, the Mesopotamians used it to pave their streets and seal the hulls of their reed boats, the Chinese found it useful for heating, and the American Indian used it for paint, fuel, and medicine. The oil they used was found on the surface, usually in pools.

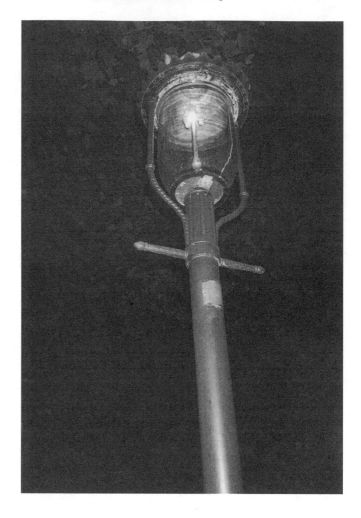

Before utility companies provided electricity to consumers, gaslight lamps were commonly used to light American streets in 1875. About 400 gaslight companies existed to provide street lighting. (Courtesy of Amy Mongillo)

The first successful oil well drilled in the United States was engineered by Edwin Drake in 1859 in Titusville, Pennsylvania. After successfully drilling for the oil, he was able to sell it for about $20 a barrel and he produced nearly 35 barrels per day. This "black gold" soon attracted many fortune seekers, but they soon discovered that technology in the petroleum industry did not keep pace with their enthusiasm. If they were fortunate enough to discover oil and put down a well, they had no feasible way to control the flow of oil. Much of the unconfined oil flowed into local streams and springs contaminating the groundwater.

The introduction of well caps and screw-like drills for deeper drilling, coupled with oil pipeline construction, made it easier to control the flow of oil. By 1865, 3.5 million barrels of petroleum per year had been pumped out of American wells. At first oil was not immediately accepted by the public; it was smelly, thick, and muddy, and it was potentially volatile. In order to be useful for home and factory it needed refining.

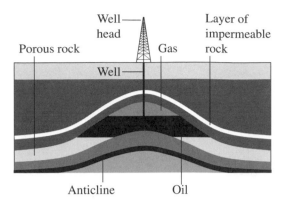

FIGURE 4-4 • **Extracting Oil** Petroleum is an important part of the United States and world economies. Petroleum is extracted from oil-bearing rocks by oil wells. An oil well is a long shaft that is drilled through rock to obtain oil. The oil is pumped up the shaft from the petroleum deposits. Many wells are drilled into an oil deposit.

Drilling for Petroleum

Although petroleum can seep to the surface, most large reserves are located deeper in the ground. To tap these reserves, drilling was necessary. In 1901 the first modern rotary rig was used at the Spindletop oil field, on a salt dome in Texas. The rotary rig used a rotary process. Attached to the rotary drill was a bit with sharp teeth which were rotated by a motor. During the drilling process, the bit cut down into the hard rock as it turned. As the drilling rigs improved, deeper wells were dug reaching deeper deposits. Many deep wells extend more than 4,000 meters (13,000 feet) deep; others reach 7,500 meters (24,000 feet) deep or more.

Oil refineries were established, and by 1860 they were producing large quantities of kerosene. This product of the refining process proved to be a very adequate substitute for whale oil for lighting. Whale oil was becoming scarce because of overhunting, and a cheaper substitute was a welcome product. Most of the byproducts left after refining the kerosene were considered waste and discarded as useless.

Chemical engineers learned to refine the crude oil so that it produced many other products. Eventually, it was refined as a gasoline, lubricant, wax, liquid petroleum gas, synthetic textiles, cosmetics, and plastics.

DID YOU KNOW?

Petroleum volume is measured in barrels. Each barrel contains approximately 159 liters (42 gallons).

Refer to Volume II for more information about petroleum and its uses.

TRANSPORTATION

Steam Power and Steam Engines

The application of steam power to industrial production and transportation was at the core of the success of the Industrial Revolution. The first steam pump was invented by Thomas Newcomen in 1712 for the express purpose of pumping water out of the deep coal mines in England. It was not very efficient or economical to run, but it worked. It encouraged men like James Watt and Richard Trevithick to explore ways in which steam power could be improved and used in other applications.

The steam engine gradually replaced waterpower as the main energy source for powering factory machinery. Mills and factories no longer needed to be located near flowing sources of water for power. They could be constructed anywhere it was appropriate for the owners and the labor force.

STEAM LOCOMOTIVES

Horse-drawn coal wagons moving along wooden tracks had been used in the mines of Great Britain for many years. Richard Trevithick's idea was to attach a steam engine on a wheeled carriage and pull coal and ore cars on an iron-edged track out of the mines. These steam carriages were called locomotives.

In 1829 George Stephenson succeeded in manufacturing a steam locomotive engine which could run at close to 24 kilometers (15 miles) per hour. Over the next two decades, his ideas were modified and improved upon by other inventors and manufacturers. Locomotives were able to carry passengers and freight more efficiently than wagons or canals. By 1850 the steam locomotive and the rail system dominated land transportation in Great Britain.

STEAMSHIPS

In 1807 Robert Fulton, using techniques he had learned in Europe, built the first commercially successful passenger steamboat. It ran on the Hudson River in New York State from New York City to Albany. Samuel Cunard ran a steamship from Liverpool, England, to Halifax, Nova Scotia, and Boston, Massachusetts, carrying mail. By the middle of the nineteenth century, regular crossings of the North Atlantic by steamship were common occurrences.

DID YOU KNOW?

An earlier locomotive run from London to Edinburgh, Scotland, for example, would take about 12 hours, far surpassing the 44 hours it took by stagecoach.

The steam locomotive, *Tom Thumb*, was built and designed in 1829 for the Baltimore & Ohio Railroad by Peter Cooper. Steam engines played an important part in railroad travel and waterway transportation systems during the 1800s. (Courtesy of the Library of Congress)

Railroads and Automobiles

RAILROADS

By 1860 the railroads were expanding rapidly throughout the country. Eventually, they became the preferred mode of transportation for moving agricultural products. They were efficient and could handle large, bulky grain products, move perishable goods to market quickly, and were not usually impeded by cold weather conditions or flood-prone rivers. The railroads were also connected to the large port cities on the eastern coastline. This meant that they could efficiently move food products from the country's interior regions to the ports for export or move foodstuffs imported from other countries to the interior regions.

AUTOMOBILES

For much of America and Europe in the nineteenth century, the railroad provided the basic source of wheeled transportation. It moved goods and people and carried most of the long-distance packages of mail, newspapers, and magazines. The railroads were highly successful in completing these tasks, but there were disadvantages to this transportation system.

Using the railroad meant that people and things moved on "railroad time." The railroad schedule determined when passengers and merchandise would leave and when they would arrive. Privacy for individuals and families was not possible when sitting with strangers, and often there were long waits between connecting train services.

In the later decades of the nineteenth century, the search had begun for a mode of transportation that would allow more personal freedom and flexibility in making local and long-haul deliveries. The invention of the automobile was thought to be a way to fulfill these goals. Several things had to be in place in order for this to be achieved. First, a reliable, durable, and affordable automobile engine and chassis had to be developed. Second, roads which had been neglected for decades had to be upgraded to handle the possibilities of long-distance travel.

Historians believe that, except for the railroad, the automobile was the most important technology for the economics of the Industrial Revolution. One of the most influential automakers was Henry Ford. In 1908–1909 he designed and produced the Model T automobile, the first mass-produced car. Ford used an *assembly line* of workers to make them. Each worker in the line performed one job. The worker added a piece to the product as it passed along to other workers in the line of work. In 1908–1909 it took the workers about 12 hours to complete each car. Ford continued to experiment with different assembly-line production strategies. By 1913, as a result of his experiments, workers could complete the production of a Model T car in less than 2 hours. As more Model Ts were built and sold, car prices came down and became more affordable to consumers. By 1927 U.S. automobile registration had reached almost 27 million motor vehicles.

The automobile became an important consumer product starting in the early 1900s. Henry Ford is standing between the first and the ten-millionth Ford. Ford's first automobile was inexpensive enough to be purchased by low and middle-income families. (Courtesy of the Library of Congress)

The booming automobile business initiated by Ford and other automakers jumpstarted other automotive-related companies which made glass windows, rubber tires, leather seats, and other automobile accessories. Steel and petroleum companies, also spurred on by the automobile business, increased their products. The construction of service stations, repair shops, and motels increased as well.

As the use of automobiles and trucks increased, highway networks were devised to connect large cities to all parts of the countryside. By 1914 there were more than 1 million cars and trucks in the United States and 250,000 in Great Britain, all moving along an improving highway system.

NEW ROADS

The condition of America's roads at the beginning of the twentieth century did not keep pace with the rapid development and use of the automobile. Roads were in a state of general neglect and disrepair. In most states the roadbed outside the urban centers was rough with uneven surfaces making travel laborious and often unsafe. As the popularity of the automobile grew, public demand for better roads grew as well.

At first, road surfaces were made up of a patchwork of dirt, brick, asphalt, and concrete. By the beginning of the twentieth century, many states had established highway departments whose purpose was to design roads that met standards of quality and workmanship that would stand up to heavy use. The use of concrete poured on several inches of crushed stone surface became the standard for most heavily used long-distance highways. Local road construction used the less expensive tar-based macadam for local roads in rural and suburban areas.

Automobile travelers and truckers took to the roads in great numbers. The Lincoln Highway and the reconditioned National Road were linked to a web of other national highways which stretched from Boston to San Francisco. The network of roads and its ever-increasing use by the public for pleasure and commercial traffic contributed to the establishment of new types of travel-related businesses.

Roadside enterprises offering food, camping, shelter, and fuel became part of the American landscape. The roads stretched the limits of the urban areas farther and farther away from the central business districts. The invention of the automobile, the extension of the highway system, and the establishment of commercial enterprises developed to support this new mode of transportation affected the lifestyle of most Americans for many decades to come.

ELECTRICITY FOR LIGHT, POWER, AND COMMUNICATIONS

Light and Power

Electricity had piqued the curiosity of scientists in the eighteenth and nineteenth centuries. Michael Farraday in England and Joseph Henry in America discovered that spinning a coil of wire through a magnetic field would produce an electric current. They also learned through experimentation that wire made of copper was an excellent conductor of electricity. This newly discovered power source was first used to produce electricity for lighthouse lamps and to electroplate gold, silver, and copper to less valuable metals.

In America Thomas Edison developed an electric generator, which produced a steady current of electricity that passed easily through copper wires. The generator was able to illuminate incandescent lightbulbs many kilometers (miles) away from the electric power plant.

Water and steam power became the primary sources for generating electricity. The water power of Niagara Falls was harnessed to turn generators. This produced electricity that passed through power lines forming a power *grid* that served customers hundreds of kilometers (miles) from the falls. Steam generators powered by burning coal could produce electricity anywhere waterpower was unavailable. By the end of the nineteenth century, the impact of electric power—its generating plants and grid of power lines—was clearly visible on the landscape.

Communications

The invention of electricity and directing it along electric wires accounted for a dramatic change in communications. Telegrams could be sent overland or by way of underground oceanic cable connecting people and continents in a matter of minutes. News of international importance, business affairs, or a personal nature could be sent from wherever a telegraph machine was located. Alexander Graham Bell's telephone (1846) connected families, friends, and businesses with the sound of the human voice. The telephone's appeal to consumers was large. In 1877 there were about 200 telephones in the United States; two years later, there were more than 56,000 telephones in use. The telephone allowed businesses and large numbers of people to communicate instantaneously.

Newspapers

The newspaper was not immune to the effects of the Industrial Revolution. In earlier times in Europe and America newspaper production was a slow, time-consuming process. Type was set by hand one letter at a time, inked, and then placed in a press that was pulled down manually over the type and paper. The newspaper was printed one quarter of a sheet at a time, and the procedure was repeated until the printing was completed. It is estimated that two men working on a press could produce, in a 10-hour day, about 200 finished newspapers.

By the 1850s, advances in giant steam-powered printing presses and new technology in newsprint production and typesetting significantly affected newspaper production. The large presses could produce 10,000 complete papers per hour; illustrations from reporters' sketches and photographs could be reproduced and included with the text. By the middle of the 1800s, over 2,500 newspapers were being published in the United States; by the 1880s, 13,000 newspapers were recorded in the national census. Before the turn of the century, newspapers in large urban areas could boast of circulation of more than 1 million per issue.

The growth in newspaper production and distribution, coupled with the development of the telegram and telephone, greatly influenced Americans and Europeans in their knowledge and perceptions of the world in which they lived.

Food and Food Distribution

The growing population in the cities and towns during the Industrial Revolution, first in Great Britain and then in the United States, created a demand for more food production and a delivery system efficient enough to get it to market. Great Britain's merchant fleet, consisting of large and quick steamships, traveled the world picking up cargo and foodstuffs and delivering it to the ports along its coastline.

Improved roads and bridges, interconnecting barge canals, and rail lines moved imported and local agricultural products to all parts of Great Britain and the United States. No one region was far from a waterway, rail terminal, or port. Products coming into port cities could be moved using any combination of transit to distribute food to the town and city markets.

HEALTH AND MEDICINE

By the middle of the nineteenth century, the ability to control infection and diagnose medical problems had been greatly improved by new technology and advances made in infectious disease research.

Philip Armour and other Chicago beef processors used refrigerated railroad cars to ship beef and other meat products to consumers all over the United States. (Courtesy of the Library of Congress)

The Rise of the Middle Class

The Industrial Revolution, accompanied by the rapid growth of cities, changed the way in which people lived together and how they related to each other. The industrial centers brought together skilled artisans and crafts-people, semiskilled and unskilled workers in the factories, building trades and transportation. It also produced a class of mostly men committed to managing the industries, establishing small businesses, improving their financial situation, and improving the quality of life in the urban setting.

These middle-class industrialists and merchants had a great influence in determining what was in their best interests and the best interests of their cities. They worked to establish an educational system that would produce a literate general population. They believed that, as the technology became more sophisticated, intelligent, literate workers would be needed to ensure the success of the economy. They wanted clean, paved roads and lighted streets, libraries, hospitals, and large halls where people could gather for lectures, theater works, and political discussions. The middle class had set the stage for people to experience a sense of community.

The discovery of X rays gave doctors faster ways to discover the extent of injuries. Vaccination techniques pioneered by Louis Pasteur helped doctors recognize and treat infectious diseases. Typhoid and cholera, two highly contagious diseases, could infect thousands in a population in a very short time. These killer diseases breed in unsanitary conditions resulting from polluted water supplies and uncontrolled public sewage. Pasteur, Robert Koch, a German physician, and other scientists began to develop vaccines to guard against these diseases and other killers such as smallpox. Programs were established for large immunization of the population, governmental laws were passed requiring a cleaner environment especially in urban areas, and health departments were established to promote and inform the public about good health habits. All of these efforts played a major role in controlling *epidemic* diseases in industrialized and crowded urban regions.

Health Conditions

The population increases in the cities and factory towns called for the construction of housing close to the factories and mines. The quality of the shelter that was constructed was generally shoddy and too small to accommodate the families that lived in them. It was reported that in the slum tenements of one city 63 people lived in 9 rooms with a bed available in each room. Small yards behind the tenements contained an outside toilet which was used by all the inhabitants of the building. Sanitary conditions were unknown.

Great masses of people were housed in the flimsy, disease-ridden tenements of the large cities. Pure water was rare; sewage ran along the ditches in the roads; and sanitary conditions were completely inadequate. Workers eventually realized that they had strength in numbers. They united to form trade unions and became active in the political process. As they became more influential and were able apply pressure to factory owners and political leaders, changes for the better occurred in housing, working conditions, and child protection.

The resulting growth of germs in these unsanitary conditions produced outbreaks of highly contagious diseases. Cholera claimed the lives of 50 percent of its victims in the first 24 hours of contact. Tuberculosis, typhus, and typhoid fever were also dangers in these unsanitary, densely populated neighborhoods.

By the late 1800s attempts were being made to control these diseases. Scientists made the connection between poor waste disposal, crowded living conditions, contaminated water supplies, and the outbreak of disease. After new sewers were built, cleaner public water was

TABLE 4-2	Discoveries in Medical Science, 1796–1900
1796	Smallpox vaccination.
1819	Stethoscope tool is invented.
1832	Hodgkin's Disease, a cancer of the lymph nodes, is described.
1842	Use of ether in surgery by Crawford W. Long.
1844	Discovery of the Cause of diphtheria.
1855	Hypodermic syringe.
1865	Tuberculosis is documented to be a contagious disease.
1879	Pasteur develops vaccines against many diseases.
1883	Koch discovers cholera as an infectious disease.
1885	Pasteur develops a vaccine against rabies.
1893	First open heart surgery is performed.
1896	X-rays are used as a diagnostic medical tool.
1897	Discovery of the Anopheles mosquito that transmits the malaria parasite.
1900	Discovery of the Aedes mosquito that spreads yellow fever.

made available, and improvements were made in housing conditions, the spread of contagious diseases was drastically reduced.

Accidents

Accidents in the mines and factories were common occurrences. In the textile mills, children working underneath the spinning machinery had an accident rate three times that of adults. In 1907 a steel mill in Pittsburgh reported that 195 men were killed in various industrial accidents in one year. In 1911 a fire broke out at the Triangle Shirtwaist Company in New York City. The fire claimed 146 workers, most of them immigrant young women. The fire was the result of fire codes being ignored. Working in the mines was especially dangerous. It is estimated that three miners were killed for every two days the anthracite coal mines were in operation. Accidents and deaths in the mines came as a result of collapsing roofs, flooding, explosions, noxious gases, runaway coal cars, and fires, as well as the long-term effects of breathing in coal dust particles: black lung disease.

Before changes were made in worker safety, statisticians in the early decades of the twentieth century estimated that nearly 25,000 workers were killed in America each year in industrial accidents, and nearly 750,000 were maimed or injured.

By the end of the nineteenth century, political pressure, public concern for the plight of the worker, and organized labor unions forced changes to take place in the working conditions in the mines and factories.

INDUSTRIAL REVOLUTION: BENEFITS AND ACHIEVEMENTS

The Industrial Revolution caused great changes in the way in which most people lived. As a result of advances made in the nineteenth and early twentieth centuries, people lived longer, were better fed and educated, could move about more quickly and safely, and could communicate far beyond the range of the human voice.

The Industrial Revolution brought about extraordinary changes on the farms and in the workplace. Machinery replaced hand labor; travel on the rails, roads, and water was safe, fast, and available to anyone with the means to pay; and communication became possible over thousands of miles over land and sea. Many new kinds of merchandise and varieties of foods became available in the marketplace.

Americans left the farms to find work in the factories and mills and swelled the population of the cities. Breakthroughs in medicine and disease control produced a healthier environment. Social issues, such as education, child labor, working conditions, and recreational opportunities, were being addressed by political leaders and social activists.

Issues of environmental concern were beginning to be addressed at the start of the twentieth century. Many of these concerns still remain with us at the beginning of the twenty-first century.

INDUSTRIAL REVOLUTION: SOCIAL, CULTURAL, AND ENVIRONMENTAL IMPACT

The Industrial Revolution resulted in great technical achievements and wealth for some members of society; however, it also bred poverty, disease, and danger for the workers who labored in its factories and mines.

The crowded living conditions in the urban centers led to the spread of filth and disease in the cramped tenements where workers lived. Working conditions in the factories and mines were dangerous and depressing. The lack of child labor laws took seven-year-old boys and girls out of school and put them into the mines and textile mills to work 12 hours a day. Millions of children under the age of 12 worked 12-hour days to help support their families.

The environment was also affected by industrial growth. The ecology of the air, land, and water was altered by a growing landscape of oil well rigging, mining, deforestation, polluted water, rail tracks, and the airborne residue of coal-fired factories.

By the end of the nineteenth century, public outrage at the conditions in which so many people lived and worked led to the establishment of labor laws, environmental controls on health and safety, and the organization of labor unions and political and social groups.

These were beginning steps in the direction of assuring all members of a society the opportunity to participate in the economic, social, and technical achievements of the Industrial Age.

TABLE 4-3	Timeline of the Industrial Revolution, 1701–1909
1701	Jethro Tull invents the seed drill freeing agricultural labor and lowering crop prices.
1709	Abraham Darby introduces coal smelting (replacing wood and charcoal) at his ironwork in England.
1712	Thomas Newcomen develops the first workable steam-powered engine pumping water out of deep coal mines.
1733	John Kay invents the Flying Shuttle greatly accelerating the weaving process and begins mechanization of the textile industry.
1756	The first coal mine in America opens in Virginia.
1759	The first Canal Act is passed by the British Parliament leading to the construction of a wide network of canals for the transport of industrial supplies and finished goods.
1763	James Hargreaves invents the spinning jenny greatly accelerating the weaving process.
1765	James Watt introduces improvements to the Newcomen steam engine.
1709	Richard Arkwright patents the water frame bringing waterpower to the textile industry.
1775	James Watt produces the first reliable, efficient steam engine.
1777	Grand Trunk Canal is opened establishing a cross-England route connecting the Mersey and Trent rivers and connecting the industrial midlands to the ports of Bristol and Liverpool.
1779	Samuel Crompton invents the spinning mule combining Hargreaves; spinning jenny and Arkwright's water frame fully mechanizing the weaving process.
1785	Edmund Cartwright invents the power loom beginning the first true mechanization of the textile industry.

1785–1799	Mass production of interchangeable parts process is developed in the arms industry in the United States led by Eli Whitney.
1786	Richard Arkwright introduces a Watt steam engine to power his cotton mill in London.
1790	Samuel Slater builds machinery in a mill in Pawtucket, Rhode Island, and operates it with waterpower.
1792	The New York Stock Exchange is established.
1793	Eli Whitney invents the cotton gin speeding up the supply of raw cotton to the textile industry.
1800	John McAdam introduces new road-making technique in Britain.
1801	Richard Trevithick demonstrates the first steam-powered locomotive.
1807	Robert Fulton begins the first regular steamboat service on the Hudson River, New York.
1811–1816	Luddites stage widespread protests in Britain against low pay and unemployment by storming factories and destroying machinery.
1821	Michael Faraday demonstrates electromagnetic rotation, the principle of the electric motor.
1825	The Erie Canal opens.
1829	Peter Cooper builds the first American locomotive.
1830	George Stephenson begins regular rail service between London and Liverpool.
1836	Samuel F. B. Morse invents the telegraph.
1837	The Great Western, the first oceangoing steamship, begins operation.
1843	Charles Thurber invents the typewriter.
1844	First commercial use of telegraph linking Baltimore and Washington.
1844	Elias Howe invents the sewing machine.
1850	First use of gasoline (refined petroleum).
1851	Isaac Singer improves on Howe's design and markets the first practical sewing machine.
1858	The "Great Stink" of London brings to the forefront the issue of rising pollution caused by industrialization for the first time.
1859	Edwin Drake drills the first oil well in Pennsylvania.
1866	Cyrus Fields lays the first transatlantic cable. The transatlantic telegraph cable is completed.
1866	Werner von Siemens perfects the dynamo to generate electricity.
1869	United States completes a transcontinental railroad.
1876	Alexander Graham Bell invents the telephone.
1877	Thomas Edison invents the phonograph.
1878	Thomas Edison invents the first practical incandescent lightbulb.
1883	First skyscraper, 10 stories high, is completed in Chicago.
1883	Brooklyn Bridge opens connecting Manhattan and Brooklyn, New York.
1886	Gottlieb Daimler and Karl Benz develop the first automobile powered by the internal combustion engine in Stuttgart, Germany.
1889	Eiffel Tower completed in Paris, the world's tallest structure until 1930.
1895	Wilhelm Roentgen discovers X rays.
1895	Guglielmo Marconi invents the wireless telegraph or radio.
1903	Orville and Wilbur Wright make the first successful airplane flight. The Ford Motor Company is formed.
1909	Henry Ford introduces the moving assembly line at his plant in Highland Park, Michigan. Ford introduces the Model T Ford.

Vocabulary

Assembly line A line of factory workers and machinery which completes a part in the production of a product and moves it to the next operation until the product is completed.

Bessemer process A process of converting pig iron into steel by decarbonizing it. Henry Bessemer's process was the most effective one at the time.

Entrepreneurs People who organize, operate, and assume the risks of a business venture.

Epidemic A contagious disease which spreads rapidly in a population.

Grid An interconnected system of cables and electric power stations spread over a wide region.

Mass production A large quantity of goods produced on an assembly line.

Activities for Students

1. Create a timeline demonstrating the evolution of transportation from the hominids to the Model T. How does this timeline compare to a timeline of the evolution of industrialization?

2. Research the effect that coal mining and black lung disease had on the health of men in Great Britain during the Industrial Revolution.

3. Contact your local environmental protection agency to find out what natural resources are indigenous to your area. How are they being depleted, protected, and replenished?

4. Contact the Centers for Disease Control and Prevention (http://www.cdc.gov; 1600 Clifton Road, Atlanta, GA 30333) to find out what vaccines are required and recommended for children and adults in the United States and in your local area.

5. Investigate the origin of child labor laws and compare the conditions to those experienced by the mill girls and breaker boys.

Books and Other Reading Materials

Adams, Samuel Hopkins. *The Erie Canal.* Eau Clair, Wisc.: Hale, 1953.

Andrist, Ralph. *Erie Canal.* New York: Dodd, American Publishing Company, 1964.

Denenberg, Barry. *So Far from Home.* New York: Scholastic, 1997.

Fisher, Leonard Everett. *The Factories.* New York: Holiday House, 1979.

Holland, Ruth. *Mill Child: The Story of Child Labor in America.* New York: MacMillan, 1970.

Macaulay, David. *Mill.* Boston: Houghton Mifflin, 1983.

———. *The Way Things Work.* Boston: Houghton Mifflin, 1988.

Paterson, Katherine. *Lyddie.* New York: Lodestar, 1990.

Rivard, Paul E. *Samuel Slater.* Pawtucket, R.I.: Slater Mill Historic Site, 1988.

Websites

Exploring World Cultures, http://evansville.net/industry

Fresno Unified School District, www.fresno.k12.ca.us

General History Articles, http://all sands.com/History

Invent Now, www.invent.org/book

Slater Mill History Museum, www.slatermill.org

Spartacus Educational, www.spartacus.schoolnet.co.uk

Steam Engine Library, www.history.rochester.edu/steam

U.S. National Archives and Records Administration, www.nara.gov/education

Economic Expansion and the Environment

The Industrial Revolution produced an *economic* revolution that was initiated by several events. During the time of the Industrial Revolution, the global population expanded rapidly. People were healthier and lived longer than in previous times. This increase in numbers provided a large labor force and a strong consumer base for new products and services.

During the Industrial Revolution, many countries established a monetary system to replace the *barter economy* in which money was scarce or did not even exist. The time was also a period of risk-taking entrepreneurs, bankers, merchants, and business groups who invested in new technologies and in expanding marketplaces on a global scale. The poor and limited transportation system was replaced by paved roads, canals, and transcontinental railroad systems. The older energy sources of power from animals, wind, and running water were superseded by new energy sources such as coal, natural gas, and petroleum.

On the plus side, the Industrial and Agricultural revolutions produced a massive expansion in economic activity throughout the world. Currently, hundreds of nations are exporting and importing large volumes of products. These products, being traded back and forth each day, include everything from automobiles, television sets, and computers to sports equipment, clothing, and food products. This widespread *economic growth* has brought social progress to many countries in the world. Food is now more accessible and affordable, *literacy* rates have improved, and life expectancy has increased. All said, a strong and growing economy has provided a high standard of living for many people.

On the minus side, however, economic growth and progress has come at a price. The natural environment has been altered and changed to meet human demands. Every day there are news reports about air and water pollution, soil erosion, the decline of plant and animal species, and environmental social injustices. If the economy continues to grow, will all of the ecosystems deteriorate eventually? Can changes be made in the present-day *economy* that will provide food, energy, and employment without further degrading the environment?

GOODS AND SERVICES

Let us look closely at some basic economic concepts to understand how an economy works and how it impacts natural resources, human resources, and energy requirements.

Goods

The production and consumption of goods and services are the driving forces behind an active economy. *Goods* can be identified as durable and nondurable products. Durable goods are those products that are usable for a relatively long period of time. They include lumber and wood products, metal products, motor vehicles, industrial machinery, and electronic equipment. Nondurable goods are those products that are usable but must be replaced within a short period of time. Some nondurable goods include food products, gasoline, clothing, paper products, beauty products, rubber and plastic products, and leather goods. Simply stated, a good is any physical object that satisfies a person's wants. In the natural environment, the goods would include food, timber, and minerals.

Services

The economic system also has *service* industries which provide services for goods. The advertising and the marketing of goods in the newspapers and on television is a service business. Auto repair mechanics provide services for automobile owners when they repair their auto-

TABLE 5-1	Nondurable Paper Goods
Paper Products	**Examples**
Containerboard	corrugated (cardboard) boxes, box dividers, reusable shipping pallets, cushioning material
Cotton Fiber Papers	fine stationery, paper money, maps, onionskin
Packaging Papers	grocery bags, yard waste bags, shopping bags, pet food bags, brown wrapping paper
Newsprint	newspaper, advertising flyers
Paperboard	milk and beverage cartons, cereal boxes, paper plates and cups, shoe boxes, board for binder covers
Printing, Writing Papers	books, magazines, copy paper, fine stationary, catalogs, direct mail pieces, envelopes, business forms, school filler paper
Specialty Papers	coffee filters, automotive filters, sandpaper, greaseproof paper, parchment paper
Tissue	bathroom tissues, facial tissue, towels, napkins

TABLE 5-2	Goods and Services
Goods	**Services**
Automobile	Auto mechanic
Refrigerator	Appliance repair
Cereal	Truck deliveries
Fruit	Packaging fruit
Bicycle	New bike advertisement
Power tool	Workshop helper
Clothing	Salesperson
Book	Publisher
Fuel	Service attendant
Cosmetics	Hairdresser
Furniture	Cabinetmaker
Bread	Baker
Baseball	Baseball player
House	Real estate agent
Toothpaste	Dentist
Medicine	Doctor
Traffic light	Police officer
Checkbook	Bank teller

mobiles. A visit to the dentist or the doctor for a checkup is a medical service. Other examples of the service industry include banking, retail and wholesale sales, real estate, marketing, and professional services in the fields of medicine, education, and government services.

In the natural environment, services include water purification, flood control, pollinization of crops, and the stabilization of climate.

The production of goods and services is essential to the standard of living for people. A growing economy creates jobs, helps reduce poverty, and provides opportunity for a better quality of life for people. The people who purchase goods and services are consumers. If the production of goods and services were to stop, the entire economy would suffer and perhaps even collapse.

THE PRODUCTION OF GOODS AND SERVICES

The production of goods and the establishment of services require natural resources, human resources, and capital goods.

Coal is an important industry for many countries. In this photo, coal is loaded into a truck from surface mining. The thickness of the coal seam can be seen to the right of the shovel. (Courtesy of Tom Repine, West Virginia Geological and Economic Survey)

Natural Resources

To produce food, fuels, furniture, clothing, housing, and metal products such as automobiles, *natural resources* are required. Examples of natural resources include air, water, soil, sunlight, minerals, ores, and fossil fuels, as well as plants and animals.

Renewable resources are those natural resources that are regularly replenished through natural processes and thus have the potential to last indefinitely. Examples of such resources include water, which is naturally replenished through the water cycle, and oxygen and carbon dioxide, which are naturally replenished through the oxygen–carbon cycle. Other renewable resources include timber, plants, animals, bacteria, fungi, and protists, which are replenished through the reproductive processes of organisms.

Nonrenewable resources are those resources that exist on Earth in fairly fixed amounts and thus have the potential of being used by organisms faster than they are replaced by nature. Some nonrenewable natural resources include minerals, ores, petroleum, natural gas, and coal.

Refer to Volume II for more information about nonrenewable and renewable resources.

Human Resources

Human resources, or the labor necessary in the production of goods or services, are also essential to the production of goods and services. Teachers, doctors, assembly-line workers, petroleum workers, office managers, custodians, real estate agents, sales people, firefighters, nurses, and police officers are all examples of human resources.

Capital Resources

One more resource is vital to produce products: *capital*. Capital provides tangible items that are made for the purpose of producing goods or services. Capital investments include tools, factories, highways, business computers, office buildings, and machines, such as tractors, railcars, and trucks. These items are needed to turn raw materials and labor into finished products which can be shipped and sold to consumers.

DIRECT COSTS

The capital needed to purchase capital goods comes from making a profit when the products are sold. A profit is the money left over after expenses for producing a product are paid. These expenses are called direct costs. The direct cost is the money spent to purchase raw materials, to supply the energy used to make the product, to buy new machinery, to build additional storage space, and to pay the labor for making the goods or services. The direct costs are passed onto the consumer in the price of the finished product.

INDIRECT COSTS

Although direct costs are associated with the production and price of a finished product, indirect costs are not. Indirect costs are those costs that are associated with the depreciation of the environment by land erosion and air and water pollution caused by the production of goods. Environmental economists believe that indirect costs should be included in the cost of the production. The company which produces the product, and/or the consumer who buys it, would pay a tax or fee which would be used to help minimize or repair environmental damage.

STAGES IN THE PRODUCTION OF GOODS AND SERVICES

The production of finished goods and services goes through several stages. In each stage, there is a relationship between the materials used in the production or service and its impact on natural resources and energy use. The major stages include the extraction of raw materials, the conversion of raw materials into manufactured goods, and the servicing of the goods. All of these stages require natural resources, human resources, and capital resources. Each stage also produces a certain amount of environmental waste and pollution.

Stage 1

The first stage includes the extraction of natural resources or raw materials. Some of these activities include drilling for petroleum, mining ores, quarrying stone, logging trees, and harvesting crops. Of all of the stages, the first stage alters the environment the most.

Stage 2

The second stage in the economic activity includes converting raw materials from the first stage, such as wood, metal, and petroleum, into finished materials. Tasks such as the distillation of petroleum to make fuels, the operation of sawmills to make lumber, and the smelting of ores to produce steel and other metals are some examples of the second stage of producing goods. Other examples include the processing of grains to make breads and other foods, the printing of newspapers and paper goods, and the converting of chemicals to make fertilizers, pesticides, pharmaceuticals, and cosmetics.

Stage 3

The last stage includes services that will be used once the finished products are produced and ready to be sold in the marketplace. Some of these services include the transportation of goods, wholesale and retail sales, internet sales, marketing and advertising, banking, financing, and real estate. All the services of the consumer goods are included in the last stage.

IMPACT OF THE THREE STAGES ON NATURAL RESOURCES

Much of the impact on the environment includes those activities such as the extraction of raw materials, the harvesting of crops, and the processing of these materials into finished goods. Of all of these, most environmentalists comment that the mining, processing, and smelting of minerals into metals have the greatest impact on the environment.

Mining

Most of the materials used in our modern economy include metals such as steel, aluminum, copper, zinc, and lead. To produce these metals, the extensive mining of ores is necessary. Mining, however, generates much waste and pollution. To extract the ore or minerals, large quantities of earth must be removed. Once the ore is separated, the remaining waste is usually piled up in heaps near the mining site. A geological survey taken in 1991 concluded that it required mining 20 million tons of ore to produce just 200,000 tons of tin. The resulting waste totaled 19.8 million tons. The extraction of gold also disfigures the natural landscape. In the same survey, to produce just 2,445 tons of gold, miners needed to extract 741 million tons of ore.

Acid Mine Drainage

In coal mining operations, the overburden waste called tailings is piled up in huge heaps. When rainwater leaches through the tailings, it causes a discharge of acidic water. The acidic water, if not treated, can flow into nearby streams and rivers polluting the water and harming aquatic life.

| TABLE 5-3 | **Metal Production and Ore Mined in Tons, 1991** |

Metal	Production	Ore Mined for Production
Iron	571,000,000	1,428,000,000
Copper	12,900,000	1,418,000,000
Lead	2,980,000	119,000,000
Aluminum	23,900,000	104,000,000
Tungsten	31,000	13,000,000

Source: U.S. Geological Survey, Department of Earth and Ocean Sciences, University of British Colombia, and Worldwatch Institute.

Acid water from coal mines is treated before releasing it into the environment. The acid water is pumped up from underground coal mines to a watery pond-like area that contains anhydrous ammonia. The ammonia reduces the iron ions in the acid water to iron (Fe^{+3}), which will precipitate out of solution. The machine in the center churns up the water to expedite the oxidation of the ions in solution. Finally the water is released and stored in a special pond nearby. (Courtesy of Tom P. Cook, West Virginia Geological and Economic Survey)

Smelting

After the ore is extracted, it is transported to a smelting operation. Smelting is the process of using heat to remove minerals or metals from ore to extract specific metals. Smelting releases sulfur dioxide emissions into the atmosphere. Copper and nickel ores contain a high percentage of sulfur. As a result, smelters of these ores release not only large amounts of sulfur dioxide into the air but they also emit nickel and copper particulate matter and other toxic chemicals. These air pollutants can contaminate the soil and damage forests as well. Smelter wastes emit gases such as carbon dioxide and carbon monoxide.

To reduce the amount of waste and pollution, responsible mining operators and smelting companies use a variety of techniques. Some mining companies use a method called reclamation—the restoring of the land to its natural setting after mining operations have been completed. Some smelting companies use scrubbers, a device that removes sulfur dioxide and other particulates from flue gases before they are emitted into the air.

Although mining operations account for much of the waste and pollution released into the environment, chemical and paper manufacturing operations are close behind. To find out more about toxic materials released into the environment in the United States, you can access the Toxic Release Inventory on the Internet.

The second stage in the production of goods produces less pollution and waste than the first stage. These wastes include cleaning products, paints, solvents, and varnishes left over in the production of goods. Other byproduct material is generated from the construction of metal products, paper goods, and clothing and fabric products. Much of this material can be recycled or reused to make finished goods thereby cutting down on the amount of waste.

In the third stage, wastes from the service activities come from consumers. These wastes include the leftover materials used in packaging products used in garden supplies, toys, and food products, as well as aluminum cans and glass containers. Some of these materials can be recycled also.

Refer to Volume IV, Human Impact on the Environment, for additional material on how humans impact their environment.

Refer to Volume V for more information about sustainable economies.

Sustainable Economy: Improving Product Quality

Currently, several automobile manufacturers are developing plans for recycling and disposing of automobile parts from old cars that are no longer usable. As an example, many different kinds of plastics are used in making automobiles: seats, the trim around windows and doors, dashboards, cupholders, some engine parts, and bumpers.

When the car is disposed of, the different plastic parts are mixed together making it difficult and costly to recycle and reprocess them. One automaker has made a change to help solve this problem. The auto maker now marks all of the different plastics that are used in the car so that someday, when it is demolished, it will be easy to identify the plastic and to sort it out appropriately. Much of the plastic used in old cars can thus be reused.

ECONOMIC MODELS

There are different economic models for the production of goods and services. Economists break down economic systems into various models in order to compare world economies. The two major ones include the free enterprise system model and the command model.

The Free Enterprise System

The economic model in the United States is known as a free market economy or market economy. The economic model is based on *capitalism*. In a capitalistic system, individuals and companies have the freedom to take part in economic activities with as little government interference as possible. In this model, the government is not responsible for producing goods and services nor do they have a planning agency to guide the economy.

America's system of capitalism goes back to colonial times in the late 1700s. In those days and now, people can make contracts, choose a variety of competitive products, and save and invest their money. They also have the right to private ownership of lands and buildings without government involvement.

As capitalism expanded in the United States, however, the government became more involved in the supervision of businesses and corporations. Laws were passed and enforced to protect markets from monopolies, to protect consumers from unsafe products, and to prevent dangerous conditions in the workplace.

The Command Model

The command model is an economic system in which the government dictates the kinds of goods that are produced. In this model, the government or state owns and controls all businesses, heavy industries such as steel and automobile manufacturing, fuel production, and banking and financing. As an example, all of the sugar production in

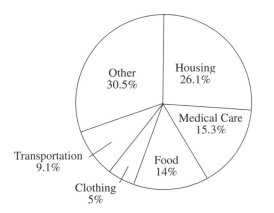

FIGURE 5-1 • Consumer Spending, 1998
Source: Economic Report of the President, 1999.

Cuba is operated and owned by the state. Cuba, China, and other communist governments use the command model. The command model also allows the government to control wages and prices. The command model in some countries may allow some free markets and private businesses to exist.

MEASURING THE ECONOMY

Gross Domestic Product

The GDP replaced the gross national product as the primary measure of U.S. production in 1991.

Economists study economic models to measure economic growth of nations. Economic growth occurs when an economy continues to produce goods and services over time. One important tool used to measure a country's economic growth or wealth is called the *gross domestic product* (GDP). The GDP is the total market value of the country's goods or services produced by labor and property during a year. In Table 5-4, the GDP is broken down to GDP per capita for several reasons.

Gross Domestic Product by Industry

The GDP calculates the value of all newly produced goods and services in an economy in a given year. In 2000, the GDP of all nations accounted for about 28 trillion dollars. Of this amount, the United States had a GDP of about 9.8 trillion dollars, followed by China with a GDP of 4.8 trillion, and Japan with about 2.5 trillion. Other countries that have a GDP between 500 billion to about 1 trillion included Italy, Brazil, India, United Kingdom, Germany, and France.

The GDP can also be broken down to GDP per capita, which reflects the average income of a country's inhabitants. In other words, it shows what part of a country's GDP each person would have if the GDP were divided and shared equally among the population. The GDP per capita does not indicate whether all people share in the wealth of a nation or whether those people lead fulfilling lives. To calculate the GDP per capita, divide a country's GDP by its total population. The GDP per capita is given in U.S. dollars. Please note that a dollar may buy more in one country than another.

Knowing a country's GDP per capita is one way of understanding a country's social, economic, and environmental strengths by measuring the general standard of living enjoyed by the inhabitants. As an example, populations in countries with a high GDP tend to have longer life expectancies, better health, higher education, better access to clean water and air, and lower infant mortality than those in the middle or at the bottom levels of the GDP. Those countries with the highest incomes per capita, however, use much of the natural and energy resources and cause the majority of global environmental problems, such as water and air pollution.

DID YOU KNOW?

About 60 percent of the people living on less than $1 a day live in South Asia and Sub-Saharan Africa.

TABLE 5-4	Gross Domestic Product per Capita, 2001

Some of the Countries above $10,000		Some of the Countries between $1,000 and $9,900		Some of the Countries at $900 or Less	
United States	$36,200	Saudi Arabia	$9,000	Yemen	$820
Singapore	$26,500	Mexico	$9,100	Madagascar	$800
Germany	$23,400	Russia	$7,700	Afghanistan	$800
United Kingdom	$21,800	Brazil	$6,500	Tanzania	$710
South Korea	$16,100	China	$3,800	Ethiopia	$600
Chile	$10,100	Egypt	$3,600	Sierra Leone	$500
		Chad	$1,000		

Source: U.S. Census Bureau.

Henry Ford

In 1896 Ford built his first automobile. It was not much more than a four-wheeled bicycle with a gas engine, but it gave him the incentive to continue working on other models. His own experiments and his application of the ideas developed by other inventors and engineers resulted in his building an assembly line where workers performed tasks on the automobile as it moved along the line. Workers added parts to the vehicle until it was completely assembled and ready for the road. This production technique made the automobile cheaper to produce and more reliable in its construction. Ford's first Model T's, which came off the assembly line in 1909, became the most popular vehicle ever produced. Its production changed the face of transportation.

Gross Domestic Product of the Sales and Production of Light Vehicles

The production, marketing, and sales of automobiles, particularly in the United States, play an important role in the global economy. In the United States, motor vehicle production accounts for approximately 3.7 percent of its GDP. The GDP of automobile production is about 126 billion dollars a year.

Researchers predict that worldwide motor vehicle ownership will increase from 600 million cars in 1999 to about 800 million by the year 2010. The annual production of automobiles in 1998 reached about 39 million. Europe manufactures about 35 percent of all automobiles in the world, followed by Asian countries (30 percent), and the United States (22 percent). The rest are produced mostly in Latin American countries, particularly Brazil. The United States and other developed countries purchase about 75 percent of all new cars. Industry experts believe that the biggest growth of automobile sales in the next decade will be in China and India. The disadvantages of too many cars on the road will include adverse effects on the environment and human health as a result of air pollution from automobile emissions.

ECONOMICS AND THE ENVIRONMENT

Environmental economists argue that economic growth does not necessarily contribute to a person's well-being if it diminishes Earth's natural resources. They contend that economic development must fit into a plan for sustainable management of resources. The sustainable economic plan focuses on improving the reduction of energy use, cutting down on product waste, and reducing consumption of energy and natural resources. Furthermore, there should be more emphasis on recycling and reusing materials.

Environmental economists are critical of the GDP because it does not take into account the losses and depreciation of the natural resources used for mining, fishing, logging, farming activities, or industrial uses. They would prefer more environmental accounting be contained in the GDP.

As an example, the accounting in a GDP statement would include a country's sale of wheat minus the costs of the depreciation of farm equipment and buildings to grow and store the wheat. However, the GDP does not adjust for any loss of natural resources such as the amount of water used to grow the crops or the depletion of fertile soil caused by soil erosion or the costs of pesticides. Nor does the GDP take in account other debits such as air and water pollution. Cleanup costs often fail to take into account cumulative or long-term environmental damage. In fact, when cleanup costs are acquired by a community, government, or business, the GDP increases. As an example, the costs of cleaning up an oil spill adds to the GDP. As a result of these concerns, members attending the Earth Summit in 1992 proposed measures to use economic and environmental accounting to provide a better indicator of a country's wealth.

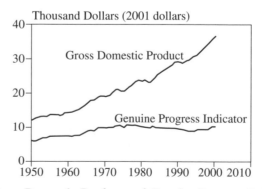

FIGURE 5-2 • Gross Domestic Product and Genuine Progress Indicator per Person, United States, 1950–2002 The Genuine Progress Indicator (GPI) was created by Redefining Progress, a U.S. nongovernment research organization. The GPI subtracts costs to the economy such as crime, pollution, and traffic. *Source:* Redefining Progress Organization.

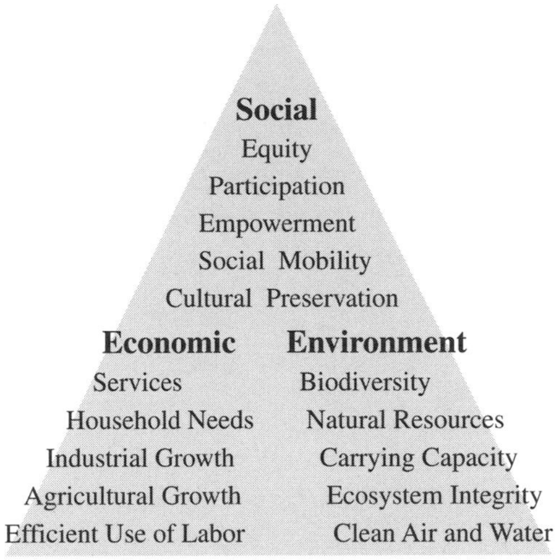

Social
Equity
Participation
Empowerment
Social Mobility
Cultural Preservation

Economic
Services
Household Needs
Industrial Growth
Agricultural Growth
Efficient Use of Labor

Environment
Biodiversity
Natural Resources
Carrying Capacity
Ecosystem Integrity
Clean Air and Water

FIGURE 5-3 • **Concept of Social, Economic, and Environmental Integration**
Source: The World Bank Group.

BUSINESS STEWARDSHIP: ECO-EFFICIENCY

Many progressive business leaders, government groups and public interest groups are looking for sustainable business practices that balance ecological, economic, and social goals. The concept of merging or integrating ecological, social, and economic goals is called eco-efficiency. The term was introduced in 1992 by the World Business Council for Sustainable Development (WBCSD). They describe eco-efficiency as a means of producing economically valuable *goods* and *services* while reducing the ecological impacts of production. The WBCSD provides business leadership which promotes eco-efficiency through high standards of environmental and resource management in business. The WBCSD comprises member companies from more than 30 nations throughout the world.

The WBCSD has identified several goals of eco-efficiency which every business should take into account when developing products. A partial list of the goals includes

- Reducing the amount of material used in the manufacturing of goods

- Becoming less dependent on fossil fuel energy

- Promoting more recyclability of goods

- Cutting down on toxic emissions and other pollutants

- Using more sustainable resources in the production of goods.

Refer to Volume V for more information on sustainable economies.

Vocabulary

Barter economy The exchange of goods and services without using money.
Capital Funds or other assets contributed to a business by the owners or stockholders.

Capitalism An economic system based on private ownership and control over the factors of production: natural resources, human resources, and capital goods. Often called a free enterprise system.

Economic The development, production, and management of material wealth as of a business.

Economic growth Occurs when an economy is able to produce more goods and services over time and is measured in gross domestic product (GDP).

Economy All of the production and consumption decisions and all activities that relate to the use of resources in a society.

Goods Any physical object, such as a car, that satisfies a person's wants.

Gross domestic product The total dollar value of all newly produced goods and services in an economy in a given year.

Literacy The ability to read and write.

Natural resources Any matter or energy derived from the environment and used by living things.

Services Any productive activity, such as a dentistry, that satisfies a person's wants.

Activities for Students

1. Draft a three-stage business plan for a product-based business of your choice. Incorporate into it all of the components identified in the "Production of Goods and Services" section of this chapter.

2. Go to the Toxic Release Inventory website and find out which toxic materials are being released into the environment in your immediate area. What regulations are in place to reduce that pollution?

3. Develop a game that demonstrates the components of the free enterprise system model and the command model.

4. Write a proposal to your city planning group for the reduction of the use of personal cars.

Books and Other Reading Materials

Brown, Lester R. *Eco-Economy, Building an Economy for the Earth.* New York: W. W. Norton, 2001.

Davies, Glyn. *A History of Money from Ancient Times to the Present Day.* Rev. ed. Cardiff: University of Wales Press, 1996.

DeCicco, John, and Martin Thomas. *The 1999 Edition of "The Green Guide to Cars and Trucks."* Washington, D.C.: American Council for an Energy-Efficient Economy.

Folsom, W. Davis. *Understanding American Business Jargon: A Dictionary.* Westport, Conn.: Greenwood Press, 1997.

Klein, Daniel B. *What Do Economists Contribute?* New York: New York University Press, 1999.

Mander, Jerry, and Edward Goldsmith, eds. *The Case against the Global Economy—and for a Turn to the Local.* San Francisco: Sierra Books, 1996.

O'Connor, David E., and Christopher Faille. *Basic Economic Principles: A Guide for Students.* Greenwood Press, Westport, Conn.: 2000.

Schumacher, E. F. *Small Is Beautiful: Economics as if People Mattered.* New York: Harper & Row, 1989.

Websites

Alliance of Automobile Manufactureres, www.autoalliance.org/

Bureau of Economic Analysis, www.bea.doc.gov/bea

Cars and Their Environmental Impact, http://www.environment.volvocars.com/ch1-1.htm

Department of Commerce, www.doc.gov/

Economic Calendar, http://biz.yahoo.com/c/terms/gdp.html

Economics Statistics Briefing Room, www.whitehouse.gov/fsbr/prices.html

Environmental Defense Fund, which provides access to data on wastes and chemicals at U.S. sources, http://www.scorecard.org

EPA National Vehicle and Fuel Emissions Laboratory, http://www.epa.gov/nvfel/

Federal Trade Commission, www.ftc.gov/

National Center for Vehicle Emissions Control and Safety (NCVECS), http://www.colostate.edu/Depts/NCVECS/ncvecs1.html

Teach with Databases, Toxic Release Inventory, http://www.nsta.org/pubs/special/pb143x01.htm

Toxic Release Inventory Public Data Release Information from the EPA, http://www.epa.gov

U.S. Environmental Protection Agency, Office of Mobile Sources, http://www.epa.gov/oms

The Growing Human Population: Food Supply and Social Issues

During the 1900s, the later stage of the Industrial Revolution, the world population was about 1.5 billion people. By the year 2000, Earth's population had reached 6 billion people. Human populations are continuing to grow at the rate of about 80 million a year. Some population experts predict that the world population will reach 8 billion by the year 2025. To feed this growing population, much demand will be placed on Earth's soil, water, and land resources to grow, harvest, transport, and distribute food. Environmentalists and *economists* are concerned about how many people Earth's resources can support.

HUMAN POPULATION

During a period of the Stone Age, the time of hunters and gatherers, the total human population was less than 1,000,000 inhabitants—less than the population of most modern cities. By A.D. 1 the population

TABLE 6-1	Population of World's 10 Largest Metropolitan Areas in 1000, 1900, and 2000					
City	**1000 (million)**	**City**	**1900 (million)**	**City**	**2000 (million)**
Cordova	0.45	London	6.5	Tokyo	26.4
Kaifeng	0.40	New York	4.2	Mexico City	18.1
Constantinople	0.30	Paris	3.3	Mumbai (Bombay)	18.1
Angkor	0.20	Berlin	2.7	São Paulo	17.8
Kyoto	0.18	Chicago	1.7	New York	16.6
Cairo	0.14	Vienna	1.7	Lagos	13.4
Bagdad	0.13	Tokyo	1.5	Los Angeles	13.1
Nishapur	0.13	St. Petersburg	1.4	Calcutta	12.9
Hasa	0.11	Manchester	1.4	Shanghai	12.9
Anhilvada	0.10	Philadelphia	1.4	Buenos Aires	12.6

Source: Molly O'Meara Sheehan, *Reinventing Cities for People and the Planet*, Worldwatch Paper 147 (Washington, D.C.: Worldwatch Institute, June 1999), pp. 14–15, with updates from United Nations, *World Urbanization Prospects: The 1999 Revision* (New York: 2000).

had reached more than 100 million people and had increased to about 500 million by 1700. During this time, most people did not live very long. Poor prenatal care, inadequate nutrition, famine, and diseases such as the *Black Death* discouraged any rapid population growth.

In the early 1800s, human population numbers grew more rapidly than during the previous three million years. By 1850 the global population reached one billion. The increased numbers were due to advancements in medicines and in food production during the Industrial Revolution and the Agricultural Revolution. By 1930 the population count was 2 billion. More food was now grown and distributed throughout the world. Many cities had built sanitation systems for waste removal and sewage treatment. Drinking water was treated and purified before reaching the public. There was also a decline in diseases such as smallpox, malaria, and cholera because of new medicines, better nutrition, and improved hospital care. All of these factors reduced the death rate.

Today, many environmentalists and others are concerned that too much growth in population will put a strain on health care, energy sources, food, water, economy, animal and plant species, and the quality of human life.

MEASURING POPULATION GROWTH AND DECLINE

Population refers to the number of organisms of the same species, such as humans, living and reproducing in a certain place. Birthrate, migration, and death rate are the major population factors studied by *demographers*.

Birthrate and Death Rate

BIRTHRATE

The birthrate is the measure of the number of births in a given year divided by the midyear population. The number of births is then multiplied by 1,000, and the sum is expressed as births per thousand population. Changes in birthrates are of concern to environmental scientists because they may indicate changes in environmental conditions. A decline in the birthrate may be an indicator of a scarcity of resources, disease, or other factors which might indicate a population or species is in decline. The birthrate is one of the factors used to determine population growth or decline. Other factors considered include *immigration*, which increases population size, and death rate and *emigration*, which decrease population size.

DEATH RATE

The death rate is the number of deaths recorded in any given year divided by the midyear population. The resulting number is multiplied by 1,000 and is expressed as deaths per thousand population. A change

TABLE 6-2	Most Populated Countries in 2001	
Rank	**Country**	**Population (millions)**
1	China	1,273
2	India	1,033
3	United States	285
4	Indonesia	206
5	Brazil	172
6	Pakistan	145
7	Russia	144
8	Bangladesh	134
9	Japan	127
10	Nigeria	127
11	Mexico	100
12	Germany	82
13	Vietnam	79
14	Philippines	77
15	Egypt	70

in the death rate, or mortality rate, is of interest to environmental scientists because it may be linked to environmental conditions. For example, scarcity of food, increased incidence of disease, and drastic climate changes, such as a drought, may all increase the death rate of a population. The death rate is one of the factors used to determine changes in population size. Other factors considered include birthrate and immigration, which increase population size, and emigration, which decreases population size.

Population Growth in Less-Developed Countries

According to the World Population Research Bureau, nearly all of the world's population growth continues to occur in *developing countries*. In these countries, such as Kenya and Bangladesh, there is a higher birthrate than death rate.

As mentioned earlier, the world population increases by about 80 million annually, and almost all of this increase occurs in the less-developed countries of Africa, Asia, and Latin America. According to current population projections, only three *developed countries*—the United States, Russia, and Japan—are expected to remain among the world's most populous by 2025. The United States is expected to remain in third place, but Russia will drop from seventh to ninth, Japan will drop from ninth to eleventh, and Germany will no longer be in the top fifteen.

TABLE 6-3	Most Populated Countries in 2025	
Rank	Country	Population (millions)
1	China	1,431
2	India	1,363
3	United States	346
4	Indonesia	272
5	Pakistan	252
6	Brazil	219
7	Nigeria	204
8	Bangladesh	181
9	Russia	137
10	Mexico	131
11	Japan	121
12	Ethiopia	118
13	Philippines	108
14	Congo, Democratic Republic of (Zaire)	106
15	Vietnam	104

Population Decline in Europe

According to the World Population Research Bureau, many European populations are experiencing more deaths than births. This reduction in birthrate is not occurring in any other world region, including the United States. Fifteen percent of Europe's population is age 65 or older, compared with 7 percent for the rest of the world.

Ukraine and Russia have the largest gaps between birthrates and death rates. The population of Ukraine is declining at a rate of 340,000 people each year from having more deaths than births. The population of Russia is declining at a rate of 950,000 people per year. In the absence of offsetting international migration, the population of these countries will continue to decline in size.

Carrying Capacity

Besides birth and death rates, demographers study the *carrying capacity* of a population. In ecology, the limit in population growth is determined by its carrying capacity. The carrying capacity can be applied to human, plant, and animal communities as well as individual species.

The carrying capacity of a human community or population is the maximum number of people a given environment can support for a period of time. As an example, if the birthrate exceeds the death rate, the population can increase beyond the carrying capacity. The increase in population will place demands on its environment in a way that will decrease future population growth; however, the population increase cannot continue for long. Usually the population growth slows down because the birthrate drops and the death rate rises. Most populations fluctuate just below and just above the carrying capacity of the environment. The carrying capacity can be displayed by two kinds of graphs that depict growth patterns of a population over time.

S-SHAPED AND J-SHAPED CURVES

If the population remains stable—that is, the birth and death rates offset each other—it is displayed as an S-shaped curve on a graph. The S-shaped curve shows a rapid growth rate followed by a slower growth rate displaying a stable population. If the population rises above the carrying capacity of the environment, the population will decline; however, the reproduction rate will increase and surpass the death rate stabilizing the population.

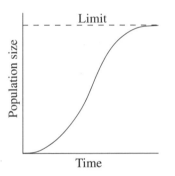

FIGURE 6-1 • S-shaped Curve A line formed on a graph showing a pattern of a successful population. The S-shaped curve shows a rapid growth rate followed by slower growth rate displaying a stable population. A stable community, one that is not threatened or becoming extinct, would be displayed with an S-shaped graph.

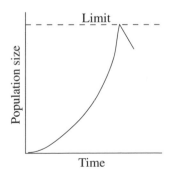

FIGURE 6-2 • J-shaped Curve A J-shaped curve plotted on a graph shows the negative growth pattern of an unsuccessful population. It displays a rise in the growth of the population and the drop point, the point at which the population is dying off. The disappearing of species would be displayed on a J-shaped graph.

A negative growth pattern, one that displays an unstable population, forms a J-shaped curve on the graph. The J-shaped curve is the plotline on the graph which swings up and then down. It displays a rise in the growth of the population and the drop point, the point at which the population is dying off. The "crash" occurs in the population when the number of individuals goes beyond the carrying capacity of a particular environment. If the reproduction rate increases and can surpass the death rate, then the population can recover; if not, the population will disappear.

How many people can Earth support? At this time, it is difficult to provide any kind of carrying-capacity models or other data that are reliable in estimating the exact number of humans Earth can support. Many variables can alter the carrying capacity of population. Some of the variables include how much food will be needed for future populations and whether birthrates will climb higher or remain stable. Another variable that affects the carrying capacity includes climate fluctuations, which can determine good or poor crop yields. Changes in the economy can alter the standards of living for many people if the economy becomes weak.

Limiting Factors

A population grows as long as enough environmental resources are available to support it; anything that holds back population growth is called a limiting factor. Limiting factors exist in all ecosystems. For example, the types and numbers of plants in an ecosystem determine the types and numbers of animals the ecosystem can support. Rainfall and temperature are examples of limiting factors. Too little or too much of either can alter an ecosystem.

The limiting factors for humans include *famine*, disease, food shortages, depletion of natural resources, social unrest, and war. *Epidemics* of disease such as yellow fever, malaria, bubonic plague, cholera, and influenza have reduced populations. Soil nutrients are the limiting factor

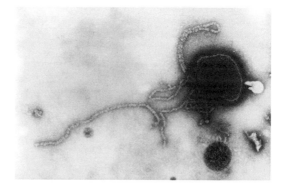

Epidemics of influenza virus have reduced populations in many countries. One of the worst global epidemics of influenza occurred in 1918–1919. According to some estimates, 20 to 40 million people, worldwide, including more than 500,000 Americans, died in this epidemic. (Courtesy of Centers for Disease Control and Prevention)

in agriculture. Older and poorer soils lack the proper nutrients, such as potassium, oxygen, carbon, sulfur, and phosphorus, to grow crops.

Droughts and wars have caused famine and starvation to millions of people in the world. As an example, during World War II, a period of about seven years, more than 60 million people perished from war-like conditions. Other limiting factors can include *natural disasters* and *climate changes*.

Limiting factors can be controlled through steps that encourage recycling, protecting and conserving natural resources, stabilizing population growth, and reducing industrialization.

Zero Population Growth

A zero population growth (ZPG) is a condition that occurs when the population size of a specific area does not increase or decrease over time. To achieve ZPG, the number of births in a population must equal the number of deaths that occur over the same period. At the same time, size increases resulting from immigration must balance with decreases resulting from emigration.

Zero population growth rarely occurs in nature; however, many people believe that the world is overpopulated with humans and think that ZPG should be a goal for human societies. Maintaining ZPG provides a means for keeping the use of resources at a constant level. In many regions of the world, human population sizes are already placing too much stress on the availability and sustainability of resources.

Analyzing and Projecting Population Trends

To forecast and project future population growth, demographers study the size, growth, and distribution of a population as mentioned earlier in the chapter. These data are provided by a census—a periodic count of a population. In the United States, the census is taken every 10 years. The data can be recorded in a population profile which shows the age structure.

The age structure is the composition of a population divided into groups by age. By studying the age structure, demographers can usually predict how the population's needs—consumption of natural resources, means of waste disposal and treatment, and need for housing, educational and health facilities, recreational space and facilities, and food supplies—are likely to change.

Age structure studies are usually performed for large areas, such as cities or entire countries. For example, a city may request an age structure study to determine the population size of school-age children for a given year in the future. Results of the study can help community officials decide how to budget funds for education. If, for example, the study predicts an increasing population, it may indicate that a new school needs to be built. Such construction may entail a variety of impacts on the environment, including the clearing of land for the

DID YOU KNOW?

Of the world's 6 billion people, more than 1.2 billion live on less than $1 a day. Two billion more people are only marginally better off.

facility, an increasing stress on water and energy resources, the need for improved wastewater treatment facilities, and the allocation of space to accommodate the production of increased solid wastes. By contrast, stable or decreasing population may indicate that moneys should be used for maintaining or improving current school buildings or to purchase textbooks or supplies. Perhaps funds should be allocated to a different societal need.

Results of age structure studies are often shown in a histogram, a bar graph that illustrates what percentages of individuals in a population fall within various age ranges. A histogram may also include a vertical axis within each bar, which divides the age data by gender.

Human Nutrition

Humans need nutrients, a substance in food needed by the body for growth, repair, and energy. Different foods supply various and different amounts of nutrients. As an example, meat, eggs, and milk provide protein as well as fat. Carbohydrates supply most of the body's energy. Foods that are high in carbohydrates include potatoes, spaghetti, and bread. Fats are a source of energy. They include meat fats as well as olive oil, corn oil, and other vegetable oils. Food also provides vitamins and minerals. These nutrients are needed for good eyesight, proper blood clotting, and disease prevention. A diet deficient in the proper amounts of nutrients can cause various diseases, such as eye disorders, retarded growth, mental disorders, weak bones, and skin disorders.

FOOD PRODUCTION AND DISTRIBUTION

Demographic studies can be used to predict the availability of food for present and future populations and their nutritional needs. Any fluctuations in food production can lead to a lack of food calories and nutrients, particularly in developing countries.

Sources of Food

Much of the food that is produced and distributed throughout the world comes from agricultural sources referred to as croplands. Much of the world's cropland is used to grow grain—wheat, rice, corn, and other *cereals*. According to the Worldwatch Institute, the average American consumes about 800 kilograms (1760 pounds) of grain a year. Some of this grain is consumed indirectly from animal products such as eggs, meat, cheese, milk, and other dairy products. Most of the food we eat is derived from land-based crops and livestock. The major grain crops include wheat, corn, rice, and, to a lesser degree, oats, barley, millet, and rye.

CROPS

CORN The United States produces about 40 percent of the world's supply of corn. In fact, the United States is the leading exporter of corn. According to the Worldwatch Institute, the United States is responsible for about 78 percent of all corn exports. Argentina, a distant second, exports about 12 percent of the world's production of corn.

WHEAT China is the world's leading producer of wheat, India is the second largest, with the United States third. Much of the wheat grown in China and India remains in those countries. The United States is the leading exporter of wheat followed by Canada, France, and Australia. The Worldwatch Institute reports that the United States, Canada, France, Australia, and Argentina account for 88 percent of the world's wheat exports. Brazil is the world's leader of wheat imports, followed by Iran, Egypt, and Japan.

RICE Besides being the leading producer of wheat, China is the leader in the production and export of rice. China is followed in rice exports by Thailand, Vietnam, and the United States.

SOYBEANS Brazil, the United States, Argentina, and China are the leading producers of soybeans. The first three countries export soybeans. China is also the largest market for U.S. soybean exports. Soybeans are legumes that fix nitrogen for corn. Corn and soybeans in the United States are rotated every two years. Rotation controls disease and the infestation of insects. Other important food crops include potatoes, sugarcane, peanuts, peas, sugar beets. Some crops such as alfalfa and hay provide food for livestock.

LIVESTOCK

A major food source includes domesticated animals or livestock. The most familiar livestock in Europe and the Americas includes pigs, sheep, goats, chickens, and cattle. Water buffalo, llamas, and even camels are sources of livestock in Asia, South America, and parts of Africa.

PORK The world's leading meat products are pork, followed by beef and poultry, which are tied for second place. China is the leading producer and consumer of pork. The country consumes about 50 percent of all global pork production.

BEEF The United States is the major producer and exporter of beef. Japan, Taiwan, and Indonesia are the leading importers of beef.

POULTRY The United States, China, France, South Africa, and Colombia are the major poultry producers. The United States accounts for 50 percent of all poultry exports.

OCEANIC FISHING AND AQUACULTURE

Approximately 5 percent of all protein in human diets comes from fish and shellfish harvested from the ocean and freshwater sources. Japan is the leading producer of oceanic fishing, followed by Russia and North America. The United States, Nova Scotia, Japan, Thailand, the Philippines, Brazil, Colombia, Australia, Portugal, and several other Mediterranean countries have large tuna industries. In aquaculture production, about 87 percent of the world's farm-raised fish is

Refer to Volume II for more information about commercial fishing and aquaculture.

Green Revolution

The Green Revolution is a period of time during which new varieties of rice and corn were developed, grown, and harvested using fertilizers and modern farming equipment. The idea behind the Green Revolution was to increase agricultural programs around the world to help reduce hunger. The Green Revolution helped farmers feed a population that almost doubled in size during a 40-year span from when it started.

The Rockefeller Foundation was the organization behind the technology that started the Green Revolution. They sponsored Norman Ernest Borlaug, a leading agricultural scientist and others to go to Mexico and help that country improve the wheat crop there. Borlaug, who began his professional career working as a forester for the U.S. Forest Service in the 1930s, later spent 15 years working as a scientist first with the Rockefeller Foundation and later for the government of Mexico. It was during this period of time that Borlaug developed several of his hardy, *hybrid* grains, which helped improve food yields in a number of developing countries throughout the world.

In the 1940s, the wheat crop in Mexico, produced very low yields, and some of the grain blew away before being harvested. By the 1960s, however, wheat production in Mexico had tripled since the arrival of Borlaug and the use of the new hybrid wheat.

Many of the plants developed by Borlaug had traits which allowed them to be grown in poor climatic conditions. The new strains also provided crops with better resistance to pests and diseases. Borlaug and other scientists developed new wheat hybrids that grew more grains per hectares (acres) when provided with the right amounts of irrigation and fertilizers. Borlaug also introduced new agricultural technologies, such as the use of machinery, pesticides, and irrigation, which increased crop production to peoples living in many developing nations. Much of Borlaug's work served the basis of the Green Revolution, and he was awarded the Nobel Peace Prize in 1970.

Later on researchers working with the Consultative Group on International Agricultural Research (CGIAR) extended Borlaug's work throughout the world. In India, wheat production tripled after the country imported the hybrid Mexican wheat in the 1960s. The new hybrids of wheat and rice also helped Latin America, Asia, China, and the Middle East produce larger crop yields.

The work done by Borlaug improved the lives of many people worldwide by providing them better access to food; however, it also introduced some new environmental problems to the areas that used the technologies and new hybrids. For example, excessive use of pesticides in some areas led to water pollution problems in nearby waterways from farmland runoff. Most of the new hybrid grains would not grow well without an irrigation system. Much of the water used for irrigation of the new crops was drawn from groundwater sources which were not sustainable. More water was discharged than recharged. The overirrigation of farmland produced waterlogged soil because of poor drainage and the accumulation of salts in the soil.

raised and harvested in Asia. Bangladesh, China, India, Indonesia, and Thailand are the leading exporters of farm-raised shrimp.

Hunger and Malnutrition

Despite the advances made in farming activities, many people still do not have enough food to eat causing malnutrition. Malnutrition is the condition of an organism that does not have all essential nutrients in the diet. The United Nations statistics show that 800 million people are chronically malnourished. Forty percent of the world's children

TABLE 6-4	**Famine Time Capsule**

1846–1847	As many as 1 million people in Ireland die from a famine caused by a potato blight. Many others emigrate from the country to avoid starvation.
1878	More than 10 million people in China are reported dead as a result of a famine brought about by two years of drought.
1928–1929	Almost 3 million people living in northern China die from famine caused by drought.
1932–1934	About 5 million people die in Russia following a famine that results when Joseph Stalin collectivizes farms and seizes rain and livestock in the Ukraine and Caucasus regions.
1941–1943	Famine resulting from World War II activities brings about the deaths of more than 40,000 people in Warsaw, Poland. During the same period, another 450,000 people living in Greece, Poland, and Yugoslavia experience famine when German soldiers block wheat shipments from the Ukraine and north Caucasus.
1943	A famine in Rwanda results in the deaths of between 35,000 and 50,000 people.
1968	A civil war in Biafra brings about a famine that kills between 0.5 and 1 million people.
1968–1974	A famine resulting from a drought in the Sahel region of Africa (located south of the Sahara) results in the deaths of 500,000 people.
1973	Famine resulting from drought is responsible for the deaths of 100,000 in Ethiopia.
1982–1984	Ethiopia is again stricken with a severe famine, this time caused by the combined effects of drought and civil war. Deaths from the famine number nearly 800,000. Another 1.5 million flee the country in search of food.

Source: John Mongillo and Linda Zierdt-Warshaw, *Encyclopedia of Environmental Science* (Westport, Conn.: Greenwood Publishing Group, 2000).

have experienced stunted growth, 33 percent are underweight, and 10 percent are wasted owing to undernourishment. As the population grows, these problems will worsen. By 2020, the number of under-nourished people may top 1 billion.

Root Causes of Hunger

A severe food shortage in an area causes widespread hunger and malnutrition. A *famine*, a limiting factor, results from both natural causes and human activities. Natural causes of famine include natural disasters such as drought, floods, earthquakes, and diseases that decimate plants and livestock. Human activities responsible for famine include areas of poverty and such events as wars, government seizures of food, the deliberate destruction of crops, and crop failures resulting from improper farming techniques leaving land unusable.

People living in developing nations are most at risk for experiencing famine, especially those resulting from natural causes. When food shortages begin, many people try to escape the problem by migrating to other areas. Such was the case in the 1840s when the potato famine of Ireland, which was caused by a fungus, resulted in one million deaths and mass migrations of that country's population to other areas.

Those unable to leave a region experiencing famine suffer from health problems associated with various forms of malnutrition. These include significant weight loss, retarded growth, dehydration, weakness, fatigue, and a diminished immune system. In time, these conditions worsen, and people die either from starvation or from diseases that spread unchecked throughout the weakened population.

To reduce the likelihood of famine, many countries have made attempts to curb population growth, or increase food production, or both. To help slow its population growth, China has mandated family planning practices in an attempt to achieve ZPG. India also has provided benefits to its people to achieve this goal.

Human Health and Disease

Improper nutrition can lead to a variety of diseases in humans. A disease is a breakdown of the body's natural defenses or regulatory systems which results in an impairment of health. Other than improper nutrition, disease can result from a variety of factors, including genetic abnormalities, pathogens (disease-causing organisms), and exposure to chemicals or other substances in the environment.

Often, an illness can be traced back to a lack of a specific nutrient in the diet. For example, kwashiorkor results from an insufficient intake of proteins, goiter from an insufficiency of salt, and rickets from an insufficiency of vitamin D. In some cases, a lack of food may lead to death by starvation, a situation affecting not only humans but many other organisms as well.

AIDS Worsening African Food Crisis

The United Nations has issued warnings over the years that acquired immune deficiency syndrome (AIDS) is worsening Africa's food crisis by killing farmers in some of the world's hungriest countries. In the past 20 years, the disease has killed 7 million farmers in Africa, cutting labor productivity by up to 50 percent, according to the joint UNAIDS agency. The agency noted that the current food crisis in southern Africa—in which an estimated 12.8 million people are at risk of starvation—is hitting countries with particularly high HIV rates.

HUMAN RIGHTS AND ENVIRONMENTAL ISSUES

There is a link between human rights and a sustainable social system and environmental problems. Environmental damage of natural resources is often worse in countries and in areas with human rights abuses. Where human rights are weak, individuals and community groups are not able to raise environmental issues and solve problems effectively. Individuals

TABLE 6-5	Income Inequality in Selected Countries, 1990s	
	Share of Income	
Country	**Poorest 20 Percent**	**Richest 20 Percent**
	(percent)	
Denmark	9.6	34.5
India	8.1	46.1
United States	5.2	46.4
Russia	4.4	53.7
Zambia	3.3	56.6
Brazil	2.2	64.1

In this table a value of zero indicates perfect equality of income. A value of 100 represents perfect inequality of income and a large economic gap between rich and poor. The list indicates that Denmark has a better rating of income equality than the other countries on the list. The growing gap between rich and poor worldwide is a major concern to environmentalists, economists, and social activists.

Source: The World Bank. Data are for the most recent year available.

need access to justice, the rights to assess information, and the freedom of expression to be successful in tackling environmental issues.

Human rights and environmental abuses are often carried out in the name of economic development. Some private companies and corporations expel local peoples from their lands, fail to put into effect pollution-control measures, and subdue local activists who question the impact of their activities on the environment.

Many poor and rural peoples in developing countries are barely informed of projects such as large dams or gas pipelines that claim to provide jobs and reduce poverty, but instead displace local inhabitants and exploit their natural resource base. Local communities, including many indigenous peoples, suffer from environmental abuses when national and state laws do not recognize their rights, particularly to their land and natural resources.

In 1994 an international group of experts on human rights and environmental protection met at the United Nations in Geneva, Switzerland. They created the Draft Declaration, the first international instrument linking principles of human rights to the environment. The following principles were included in the Draft Declaration:

- All persons have the right to a secure, healthy, and ecologically sound environment.

- All persons shall be free from any form of discrimination in regard to actions and decisions that affect the environment.

- All persons have the right to freedom from pollution, environmental degradation and activities that adversely affect the environment, threaten life, and health.

- All persons have the right to safe and healthy food and water adequate to their well-being.

- All persons have the right to a safe and healthy working environment.

- All persons have the right to adequate housing, land tenure and living conditions in a secure, healthy and ecologically sound environment.

Additional human rights violations and abuses include ethnic persecution and environmental racism. In 1987 the Reverend Benjamin Franklin Chavis, Jr., used the term "environmental racism" to describe the unfair distribution of dumps and incinerators in minority neighborhoods. The term was used in a study conducted by the United Church of Christ Commission for Racial Justice under the leadership of Reverend Chavis. The study defined environmental racism as racial discrimination in environmental policy making, the deliberate targeting of communities of color for toxic waste facilities, and the official sanctioning of the life-threatening presence of poisons and pollutants in minority communities. Concerned that minority populations and low-income populations were bearing a disproportionate amount of adverse health and environmental effects, President William J. Clinton issued an Executive Order in 1994, focusing federal agency attention on these issues. The Environmental Protection Agency (EPA) responded by developing the Environmental Justice Strategy which focuses on addressing these concerns.

A sustainable environment and economy will require social systems that are just and equitable. These strategies include giving local communities more authority and control over their own resources, allowing them access to freshwater and sanitation, better nutrition, job training and education, and the opportunity to create their own wealth.

Vocabulary

Black Death An epidemic disease, also known as the bubonic plague, which spread through much of Europe during the fourteenth century. The bacterial disease was caused by the spread of fleas that lived on rodents.

Carrying capacity Maximum number of individuals that can be supported by an environment.

Cereals Specialized types of grasses which are cultivated for grain.

Climate changes Change in the annual temperature over a period of time.

Demographers People who study populations.

Developed countries Countries which are highly industrialized whose people have high incomes per capita (for each person).

Developing countries Countries that are not fully industrialized whose people have low per capita (for each person) incomes. The major economies are agricultural. Approximately 70 percent of the world's population live in developing countries.

Economists People who study the economy and the choices made by individuals, governments, and businesses in that economy.

Emigration Leaving one country.

Epidemic A disease that spreads quickly through a large population.

Famine A period of severe shortages of food.

Hybrid A cross between two species or varieties of plants or animals to increase growth and resistance to disease.

Immigration Coming into a new country.

Natural Disaster Change in the environment, such as an earthquake, that is not the result of human actions.

Activities for Students

1. Create a series of overlay maps that outline population expansion over time, from the Stone Age to the modern day. Include major cities and civilizations.

2. Find the birth and death rates of three countries from a first-, second- and third-world economy. Write about what you notice and the possible causes of the data.

3. Visit or contact the Census Bureau. What is the population growth pattern for the next 10 years? Where will the major expansions occur? What infrastructure needs to be in place to support the traffic, education, and waste management of your town?

4. Research on the Website "The Real Reasons for Hunger," by Vandana Shiva, and write a response article. www.globalpolicy.org/globaliz/econ/2002/0623hunger.htm

Books and Other Reading Materials

Bartoletti, Susan Campbell. *Black Potatoes: The Story of the Great Irish Famine, 1845–1850*. Boston: Houghton Mifflin, 2001.

Bullard, Robert D. *Confronting Environmental Racism*. Boston: South End Press, 1993.

Chiras, Daniel D. *Environmental Science: A Framework for Decision Making*. Menlo Park, Calif.: Addison-Wesley, 1989.

Cohen, Joel E. *How Many People Can the Earth Support?* New York: W. W. Norton, 1995.

Ehrlich, Paul. *The Population Explosion*. New York: Touchstone Books, 1991.

Fagan, Brian M. *Floods, Famines, and Emperors: El Niño and the Fate of Civilizations*. New York: Basic Books, 1999.

Saw, Swee-Hock. *Population Control for Zero Growth in Singapore*. Cambridge, England: Oxford University Press, 1980.

U.S. Census Bureau. *2000 America! U.S. Census Bureau National Facts, Statistics, and Population Data*. CD-ROM Progressive Management, 2000.

Welch, R. M. *Crops as Sources of Nutrients for Humans*. ASA Special Publications, no. 48. Madison, Wis.: American Society of Agronomy, 1984.

General Websites

PopPlanet (Population, Health, and Environment), http://PopPlanet.org/

Population Reference Bureau, http://www. prb.org/

Real Reasons for Hunger, www.globalpolicy.org/globaliz/econ/2002/0623hunger.htm

Zero Population Growth, http://www.zpg.org

Appendix A: Environmental Timeline, 1620–2004

Environmentalists and activists appear in **boldface**.

1620 to 1860 Erosion becomes a major problem on many American farms. Fields are abandoned. Rivers and streams are filled with silt and mud. The publication of farm journals is initiated by early soil conservationists to improve farming methods.

1748 Jared Eliot, a minister and doctor of Killingsworth, Connecticut, writes the first American book on agriculture to improve crops and to conserve soil.

1824 Solomon and William Drown of Providence, Rhode Island, publish *Farmer's Guide* which discusses erosion and its causes and remedies. A year later, John Lorain, of the Philadelphia Agricultural Society, publishes a book devoted to the prevention of soil erosion in which he discusses methods such as using grass as an erosion-control crop.

1827 John James Audubon begins publication of *Birds of America*.

1830 George Catlin launches his great western painting crusade to document Native American peoples.

1845 Henry David Thoreau moves to Walden Pond to observe the fauna and flora of Concord, Massachusetts.

1847 U.S. Congressman **George Perkins Marsh** of Vermont delivers a speech calling attention to the destructive impact of human activity on the land.

1849 The U.S. Department of the Interior (DOI) is established.

1857 Frederick Law Olmsted develops the first city park: New York City's Central Park.

1859 British naturalist Charles Darwin publishes *The Origin of the Species by Means of Natural Selection*. In time the theory of evolution presented in the book becomes the most widely accepted theory of evolution.

1866 German biologist Ernst Haeckel introduces the term *ecology*.

1869 John Muir moves to the Yosemite Valley.

Geologist and explorer John Wesley Powell travels the Colorado River through the Grand Canyon.

1872 Yellowstone National Park is established as the first national park of the United States in Yellowstone, Wyoming.

U.S. legislation: Passage of the Mining Law permits individuals to purchase rights to mine public lands.

1876 The Appalachian Mountain Club is founded.

1879 The U.S. Geological Survey (USGS) is formed.

1882 The first hydroelectric plant opens on the Fox River in Wisconsin.

1883 Krakatoa, a small island of Indonesia, is virtually destroyed by a volcanic explosion.

1890 Denmark constructs the first windmill for use in generating electricity.

Sequoia National Park, Yosemite National Park, and General Grant National Park are established in California.

1891 U.S. legislation: Passage of Forest Reserve Act provides the basis for a system of national forests.

1892 John Muir, Robert Underwood Johnson, and William Colby are cofounders of the Sierra Club, in Muir's words, to "do something for wildness and make the mountains glad."

1893 The National Trust is founded in the United Kingdom. The group purchases land deemed of having natural beauty or considered a cultural landmark.

1895 Founding of the American Scenic and Historic Preservation Society.

1898 Cornell University establishes the first college program in forestry.

Gifford Pinchot becomes head of the U.S. Division of Forestry (now the U.S. Forest Service) and serves until 1910. Under President

Theodore Roosevelt, many of Pinchot's ideas became national policy. During his service, the national forests increase from 32 in 1898 to 149 in 1910, a total of 193 million acres.

1899 The River and Harbor Act bans pollution of all navigable waterways. Under the act, the building of any wharves, piers, jetties, and other structures is prohibited without congressional approval.

1900 U.S. legislation: Passage of Lacey Act makes it unlawful to transport illegally killed game animals across state boundaries.

1902 U.S. legislation: Passage of Reclamation Act establishes the Bureau of Reclamation.

1903 First federal U.S. wildlife refuge is established on Pelican Island in Florida.

1905 The National Audubon Society, named for wildlife artist John James Audubon, is founded.

1906 Yosemite Valley is incorporated into Yosemite National Park.

1907 International Association for the Prevention of Smoke is founded. The group's name later changes several times to reflect other concerns over causes of air pollution.

Gifford Pinchot is appointed the first chief of the U.S. Forest Service.

1908 The Grand Canyon is set aside as a national monument.

Chlorination is first used at U.S. water treatment plants.

President Theodore Roosevelt hosts the first Governors' Conference on Conservation.

1914 The last passenger pigeon, Martha, dies in the Cincinnati zoo.

1916 The National Park Service (NPS) is established.

1918 Hunting of migratory bird species is restricted through passage of the Migratory Bird Treaty Act. The act supports treaties between the United States and surrounding nations.

Save-the-Redwoods League is created.

1920 U.S. legislation: Passage of the Mineral Leasing Act regulates mining on federal lands.

1922 The Izaak Walton League is organized under the direction of **Will H. Dilg**.

1924 Environmentalist **Aldo Leopold** wins designation of Gila National Forest, New Mexico, as first extensive wilderness area.

Marjory Stoneman Douglas, of the *Miami Herald*, writes newspaper columns opposing the draining of the Florida Everglades.

Bryce Canyon National Park is established in Utah.

1925 The Geneva Protocol is signed by numerous countries as a means of stopping use of biological weapons.

1928 The Boulder Canyon Project (Hoover Dam) is authorized to provide irrigation, electric power, and a flood-control system for Arizona and Nevada communities.

1930 Chlorofluorocarbons (CFCs) are deemed safe for use in refrigerators and air conditioners.

1931 France builds and makes use of the first Darrieus aerogenerator to produce electricity from wind energy.

Addo Elephant National Park is established in the Eastern Cape region of South Africa to provide a protected habitat for African elephants.

1932 Hugh Bennett is given the opportunity to put his soil conservation ideas into practice to help reduce soil erosion. He becomes the director of the Soil Erosion Service (SES) created by the Department of Interior.

1933 The Tennessee Valley Authority (TVA) is formed.

The Civilian Conservation Corps (CCC) employs more than 2 million Americans in forestry, flood control, soil erosion, and beautification projects.

1934 The greatest drought in U.S. history continues. Portions of Texas, Oklahoma, Arkansas, and several other midwestern states are known as the "Dust Bowl."

U.S. legislation: Passage of Taylor Grazing Act regulates livestock grazing on federal lands.

1935 The Soil Conservation Service (SCS) is established.

The Wilderness Society is founded.

1936 The National Wildlife Federation (NWF) is formed.

1939 David Brower produces his first nature film for the Sierra Club, called *Sky Land Trails of the Kings*. In the same year, Brower, who is an excellent climber, completes his most famous ascent, Shiprock, a volcanic plug which rises 1,400 feet from the floor of the New Mexico desert.

1940 The U.S. Wildlife Service is established to protect fish and wildlife.

U.S. legislation: President Franklin Roosevelt signs the Bald Eagle Protection Act.

1945 The United Nations (UN) establishes the Food and Agriculture Organization (FAO).

1946 The International Whaling Commission (IWC) is formed to research whale populations.

The U.S. Bureau of Land Management (BLM) and the Atomic Energy Commission (AEC) are created.

1947 Marjory Stoneman Douglas publishes *The Everglades: River of Grass* and serves as a member of the committee that gets the Everglades designated a national park.

1948 The UN creates the International Union for the Conservation of Nature (IUCN) as a special environmental agency.

An air pollution incident in Donora, Pennsylvania, kills 20 people; 14,000 become ill.

U.S. legislation: Passage of Federal Water Pollution Control Law.

1949 Aldo Leopold's *A Sand County Almanac* is published posthumously.

1950 Oceanographer **Jacques Cousteau** purchases and transforms a former minesweeper, the *Calypso*, into a research vessel which he uses to increase awareness of the ocean environment.

1951 Tanzania begins its national park system with the establishment of the Serengeti National Park.

1952 Clean air legislation is enacted in Great Britain after air pollution–induced smog brings about the deaths of nearly 4,000 people.

David Brower becomes the first executive director of the Sierra Club.

1953 Radioactive iodine from atomic bomb testing is found in the thyroid glands of children living in Utah.

1955 U.S. legislation: Passage of the Air Pollution Control Act, the first federal legislation designed to control air pollution.

1956 U.S. legislation: Passage of the Water Pollution Control Act authorizes development of water-treatment plants.

1959 The Antarctic Treaty is signed to preserve natural resources of the continent.

1961 The African Wildlife Foundation (AWF) is established as an international organization to protect African wildlife.

1962 **Rachel Carson** publishes *Silent Spring*, a groundbreaking study of the dangers of DDT and other insecticides.

Hazel Wolf joins the National Audubon Society in Seattle, Washington, and plays a prominent role in local, national, and international environmental efforts during her lifetime.

1963 The Nuclear Test Ban Treaty between the United States and the Soviet Union stops atmospheric testing of nuclear weapons.

U.S. legislation: Passage of the first Clean Air Act (CAA) authorizes money for air pollution control efforts.

1964 **Hazel Henderson** organizes women in a local play park in New York City and starts a group called Citizens for Clean Air, the first environmental group, she believes, east of the Mississippi. She built Citizens for Clean Air from a very small group to a membership of 40,000. Two years later, 80 people died in New York City from air pollution–related causes during four days of atmospheric inversion.

U.S. legislation: Passage of the Wilderness Act creates the National Wilderness Preservation System.

1965 U.S. legislation: Passage of the Water Quality Act authorizes the federal government to set water standards in absence of state action.

1966 Eighty people in New York City die from air pollution–related causes.

1967 The *Torey Canyon* runs aground spilling 175 tons of crude oil off Cornwall, England.

Dian Fossey establishes the Karisoke Research Center in the Virunga Mountains, within the Parc National des Volcans in Rwanda to study endangered mountain gorillas.

The Environmental Defense Fund (EDF) is formed to lead an effort to save the osprey from DDT.

1968 U.S. legislation: Passage of the Wild and Scenic Rivers Act and the National Trails System Act identify areas of great scenic beauty for preservation and recreation.

Paul Ehrlich publishes *The Population Bomb*.

1969 Wildlife photographer Joy Adamson establishes the Elsa Wild Animal Appeal, an organization dedicated to the preservation and humane treatment of wild and captive animals.

Greenpeace is created.

Blowout of oil well in Santa Barbara, California, releases 2,700 tons of crude oil into the Pacific Ocean.

U.S. legislation: Passage of the National Environmental Policy Act (NEPA) requires all federal agencies to complete an environmental impact statement for any dam, highway, or other large construction project undertaken, regulated, or funded by the federal government.

The Friends of the Earth (FOE) is founded in the United States.

John Todd, **Nancy Jack Todd**, and Bill McLarney are the cofounders of the New Alchemy Institute in Cape Cod, Massachusetts. The institute begins to pioneer a new way of treating sewage and other wastes.

1970 Denis Hayes is the national coordinator of the first Earth Day, which is celebrated on April 22.

Construction of the Aswan High Dam on the Nile River in Egypt is completed.

U.S. legislation: Passage of an amended Clean Air Act (CAA) expands air pollution control.

The U.S. Environmental Protection Agency (EPA) is established.

1971 Canadian primatologist Biruté Galdikas begins her studies of orangutans through the Orangutan Research and Conservation Project in Borneo.

The United Nations Educational, Scientific and Cultural Organization (UNESCO) establishes the Man and the Biosphere Program, developing a global network of biosphere reserves.

1972 The Biological and Toxin Weapons Convention is adopted by 140 nations to stop the use of biological weapons.

The EPA phases out the use of DDT in the United States to protect several species of predatory birds. The ban builds on information obtained from Rachel Carson's 1962 book, *Silent Spring*.

U.S. legislation: Passage of the Water Pollution Control Act, the Coastal Zone Management Act (CZMA), and the Environmental Pesticide Control Act.

Oregon passes the first bottle-recycling law.

1973 Norwegian philosopher Arne Naess coins the term *deep ecology* to describe his belief that humans need to recognize natural things for their intrinsic value, rather than just for their value to humans.

The Convention on International Trade in Endangered Species of Wild Fauna and Flora (CITES) is signed by more than 80 nations. The Endangered Species Act of the United States also is enacted.

Congress approves construction of the 1,300-kilometer pipeline from Alaska's North Slope oil field to the Port of Valdez.

An Energy crisis in the United States arises from an Arab oil embargo.

A collision and resulting explosion between the *Corinthos* oil tanker and the *Edgar M. Queeny* releases 272,000 barrels of crude oil and other chemicals into the Delaware River near Marcus Hook, Pennsylvania.

1974 Scientists report their discovery of a hole in the ozone layer above Antarctica.

U.S. legislation: Passage of the Safe Drinking Water Act sets standards to protect the nation's drinking water. The EPA bans most uses for aldrin and dieldrin and disallows the production and importation of these chemicals into the United States.

1975 Unleaded gas goes on sale. New cars are equipped with antipollution catalytic converters.

The EPA bans use of asbestos insulation in new buildings.

Edward Abbey publishes *The Monkey Wrench Gang*, a novel detailing acts of ecotage as a means of protecting the environment.

1976 *Argo Merchant* runs aground releasing 25,000 tons of fuel into the Atlantic Ocean near Nantucket, Rhode Island.

National Academy of Sciences reports that CFC gases from spray cans are damaging the ozone layer.

U.S. legislation: Passage of the Resource Conservation and Recovery Act empowers the EPA to regulate the disposal and treatment of municipal solid and hazardous wastes. The Toxic Substances Control Act and the Resource Conservation and Recovery Act are enacted.

Fire aboard the *Hawaiian Patriot* releases nearly 100,000 tons of crude oil into the Pacific Ocean.

1977 The Green Belt Movement is begun by Kenyan conservationist Wangari Muta Maathai on World Environment Day.

Blowout of Ekofisk oil well releases 27,000 tons of crude oil into the North Sea.

Construction of the Alaska pipeline, the 1,300-kilometer pipeline that carries oil from

Alaska's North Slope oil field to the Port of Valdez, is completed at a cost of more than $8 billion.

U.S. legislation: Passage of the Surface Mining Control and Reclamation Act.

The Department of Energy (DOE) is created.

1978 The *Amoco Cadiz* tanker runs aground spilling 226,000 tons of oil into the ocean near Portsall, Brittany.

People living in the Love Canal community of New York are evacuated from the area to reduce their exposure to chemical wastes which have surfaced from a canal formerly used as a dump site.

Rainfall in Wheeling, West Virginia, is measured at a pH of 2, the most acidic rain yet recorded.

Aerosols with fluorocarbons are banned in the United States.

The EPA bans the use of asbestos in insulation, fireproofing, or decorative materials.

1979 British scientist **James E. Lovelock** publishes *Gaia: A New Look at Life on Earth.*

Collision of the *Atlantic Empress* and the *Aegean Captain* releases 370,000 tons of oil into the Caribbean Sea.

The Convention on Long-Range Transboundary Air Pollution (LRTAP) is signed by several European nations to limit sulfur dioxide emissions which cause acid rain problems in other countries.

The Three Mile Island Nuclear Power Plant in Pennsylvania experiences near-meltdown.

The EPA begins a program to assist states in removing flaking asbestos insulation from pipes and ceilings in school buildings throughout the United States.

The EPA bans the marketing of herbicide Agent Orange in the United States.

1980 Debt-for-nature swap idea is proposed by Thomas E. Lovejoy: nations could convert debt to cash which would then be used to purchase parcels of tropical rain forest to be managed by local conservation groups.

Global Report to the President addresses world trends in population growth, natural resource use, and the environment by the end of the century, and calls for international cooperation in solving problems.

U.S. legislation: Passage of the Comprehensive Environmental Response, Compensation, and Liability Act (Superfund) and the Low Level Radioactive Waste Policy Act.

1981 Earth First!, a radical environmental action group that resorts to ecotage to gain its objectives, formed.

Lois Gibbs founds the Citizens' Clearinghouse for Hazardous Wastes, later named the Center for Health, Environment, and Justice (CHEJ).

1982 U.S. legislation: Passage of the Nuclear Waste Policy Act.

1983 A film of **Randy Hayes**, *The Four Corners, a National Sacrifice Area*, wins the 1983 Student Academy Award for the best documentary. The film documents the tragic effects of uranium and coal mining on Hopi and Navajo Indian lands in the American Southwest.

The residents of Times Beach, Missouri, are ordered to evacuate their community. Investigations of Times Beach in the 1980s disclosed the fact that oil contaminated with dioxin, a highly toxic substance, had been used to treat the town's streets.

Cathrine Sneed founds and acts as director of the Garden Project in San Francisco. The Garden Project, a horticulture class for inmates of the San Francisco County Jail, uses organic gardening as a metaphor for life change. The U.S. Department of Agriculture calls the project "one of the most innovative and successful community-based crime prevention programs in the country."

1984 Toxic gases released from the Union Carbide chemical manufacturing plant in Bhopal, India kill an estimated 3,000 people and injure thousands of others.

The Jane Goodall Institute (JGI) is founded.

The British tanker *Alvenus* spills 0.8 million gallons of oil into the Gulf of Mexico.

U.S. legislation: Passage of the Hazardous and Solid Waste Amendments.

1985 Concerned Citizens of South Central Los Angeles becomes one of the first African American environmental groups in the United States. **Julia Tate** serves as the executive director. The organization's goal is to provide a better quality of life for the residents of this Los Angeles community. **Maria Perez**, **Nevada Dove**, and **Fabiola Tostado** later join the group and are known as the Toxic Crusaders.

Huey D. Johnson becomes the founder and president of the Resource Renewal Institute

(RRI), a nonprofit organization located in California. Johnson suggests that green plans is the path countries should take to respond to environmental decline. Green plans treat the environment as it really exists—a single, interconnected ecosystem that can be safeguarded for future generations only through a systemic, long-range plan of action.

Scientists of the British Antarctica Survey discover the ozone hole. The hole, which appears during the Antarctic spring, is caused by the chlorine from CFCs.

Juana Gutiérrez becomes president and founder of Mothers of East Los Angeles, Santa Isabel Chapter (Madres del Este de Los Angeles—Santa Isabel) (MELASI) whose mission is to fight against toxic dumps and incinerators and also to take a proactive approach to community improvement.

Primatologist Dian Fossey is discovered murdered in her cabin at the Karosoke Research Center she founded. Her death is attributed to poachers.

While protesting nuclear testing being conducted by France in the Pacific Ocean, the *Rainbow Warrior* (a boat owned by Greenpeace) is sunk in a New Zealand harbor by agents of the French government.

U.S. legislation: Passage of the Food Security Act.

1986 Tons of toxic chemicals stored in a warehouse owned by the Sandoz pharmaceutical company are released into the Rhine River near Basel, Switzerland. The effects of the spill are experienced in Switzerland, France, Germany, and Luxembourg.

An explosion destroys a nuclear power plant in Chernobyl, Ukraine, immediately killing more than 30 people and leading to the permanent evacuations of more than 100,000 others.

Bovine spongiform encephalopathy (BSE), a neurodegenerative illness of cattle, also known as mad cow disease, comes to the attention of the scientific community when it appears in cattle in the United Kingdom.

U.S. legislation: Passage of the Emergency Response and Community Right-to-Know Act and the Superfund Amendments and Reauthorization Act (SARA).

1987 The Montreal Protocol, an international treaty that proposes to cut in half the production and use of CFCs, is approved by more than 30 nations.

The world's fourth largest lake, the Aral Sea of Asia, is divided in two as a result of the diversion of water from its feeder streams, the Syr Darya and Amu Darya rivers.

The *Mobro*, a garbage barge from Long Island, New York, travels 9,600 kilometers in search of a place to offload the garbage it carries.

1988 Use of ruminant proteins in the preparation of cattle feed is banned in the United Kingdom to prevent outbreaks of BSE.

Global temperatures reach their highest levels in 130 years.

The Ocean Dumping Ban legislates international dumping of wastes in the ocean.

EPA studies report that indoor air can be 100 times as polluted as outdoor air. Radon is found to be widespread in U.S. homes.

Beaches on the east coast of the United States are closed because of contamination by medical waste washed onshore.

The United States experiences its worst drought in 50 years.

Plastic ring six-pack holders are required to be made degradable.

U.S. legislation: Passage of the Plastic Pollution Research and Control Act bans ocean dumping of plastic materials.

1989 The United Kingdom bans the use of cattle brains, spinal cords, tonsils, thymuses, spleens, and intestines in foods intended for human consumption as a means of preventing further outbreaks of Creutzfeldt-Jakob disease (CJD), the human version of mad cow disease, in humans.

Fire aboard the *Kharg 5* releases 75,000 tons of oil into the sea surrounding the Canary Islands.

The Montreal Protocol treaty is updated and amended.

The New York Department of Environmental Conservation reports that 25 percent of the lakes and ponds in the Adirondacks are too acidic to support fish.

The *Exxon Valdez* runs aground on Prince William Sound, Alaska, spilling 11 million gallons of oil into one of the world's most fragile ecosystems.

1990 Ocean Robbins, age 16, and **Ryan Eliason**, 18, are the cofounders of YES!, or Youth for Environmental Sanity. The goal of YES! is to educate, inspire, and empower young people to take positive action for healthy people and a healthy planet. Robbins served as director for five years and is now the organization's president. As of 2000, the program has reached 600,000 students in 1,200 schools in 43 states through full school assemblies.

UN report forecasts a world temperature increase of 2°F within 35 years as a result of greenhouse gas emissions.

U.S. legislation: Passage of the Clean Air Act amendments including requirements to control the emission of sulfur dioxide and nitrogen oxides.

1991 The Gulf War concludes with hundreds of oil wells in Kuwait being set afire by Iraqi troops, resulting in extensive air and water pollution problems.

The United States accepts an agreement on Antarctica which prohibits activities relating to mining, protects native species of flora and fauna, and limits tourism and marine pollution.

Eight scientists begin a two-year stay in Biosphere 2 in Arizona, a test center designed to provide a self-sustaining habitat modeling Earth's natural environments. The experiment, which is repeated in 1993, meets with much criticism and is deemed largely unsuccessful.

1992 UN Earth Summit is held in Rio de Janeiro, Brazil. Major resolutions resulting from the summit include the Rio Declaration on Environment and Development, Agenda 21, Biodiversity Convention, Statement of Forest Principles, and the Global Warming Convention, which is signed by more than 160 nations.

Severn Cullis-Suzuki speaks for six minutes to the delegates urging them to work hard on resolving global environmental issues. She received a standing ovation.

The Montreal Protocol is again amended with signatories agreeing to phase out CFC use by the year 2000.

1993 Sugar producers and U.S. government agree on a restoration plan for the Florida Everglades.

1994 *Dumping in Dixie: Class and Environmental Quality* is published by **Robert Bullard**. The book reports on five environmental justice campaigns in states ranging from Texas to West Virginia. Bullard emphasizes that African Americans are concerned about and do participate in environmental issues.

The California Desert Protection Act is passed.

Failure of a dike results in the release of 102,000 tons of oil into the Siberian tundra near Usink in northern Russia.

The Russian government calls for preventive measures to control the destruction of Lake Baikail.

The bald eagle is reclassified from an endangered species to a threatened species on the U.S. Endangered Species List.

An 8.5-million-gallon spill is discovered in Unocal's Guadalupe oil field in California.

1995 The U.S. Government reintroduces endangered wolves to Yellowstone Park.

1999 Scientists report that the human population of Earth now exceeds 6 billion people.

The peregrine falcon is removed from the U.S. Endangered Species List.

The *New Carissa* runs aground off the coast of Oregon, leaking some oil into Coos Bay. The tanker is later towed into the open ocean and sunk.

Beyond Globalization: Shaping a Sustainable Global Economy is published by Hazel Henderson.

Paul Hawken coauthors *Natural Capitalism, Creating the Next Industrial Revolution.*

Off the Map, an Expedition Deep into Imperialism, the Global Economy, and Other Earthly Whereabouts is published by **Chellis Glendinning**.

Twenty-three-year-old **Julia Butterfly Hill** comes down out of a 180-foot California redwood tree after living there for two years to prevent the destruction of the forest. A deal is made with the logging company to spare the tree as well as a three-acre buffer zone.

2000 Denis Hayes is the coordinator and **Mark Dubois** is the international coordinator of Earthday 2000.

Ralph Nader and **Winona LaDuke** run for U.S. president and vice president on the Green Party ticket.

In January 2000, Hazel Wolf passes away at the age of 101.

The Chernobyl nuclear power plant is scheduled to close in December.

Anthropologists for the Wildlife Conservation Society in New York announce that a type of large West African monkey is extinct, making it the first primate to vanish in the twenty-first century.

A study by National Park Trust, a privately funded land conservancy, finds that more than 90,000 acres within state parks in 32 states are threatened by commercial and residential development and increased traffic, among other things.

A bone-dry summer in north-central Texas breaks the Depression-era drought record when the Dallas area marks 59 days without rain. The arid streak with 100-degree daily highs breaks a record of 58 days set in the midst of the Dust Bowl in 1934 and tied in 1950. The Texas drought exceeded 1 billion dollars in agricultural losses.

Massachusetts announces that the state will spend $600,000 to determine whether petroleum pollution in largely African American city neighborhoods contributes to lupus, a potentially deadly immune disease. The research, to be conducted over three years, will target three areas of the city with unusually high levels of petroleum contamination.

Hybrid vehicle Toyota Prius is offered for sale in the United States.

The hole in the ozone layer over Antarctica has stretched over a populated city for the first time, after ballooning to a new record size. Previously, the hole had opened only over Antarctica and the surrounding ocean.

2001 An environmental group that successfully campaigned for the return of wolves to Yellowstone National Park wants the federal government to do the same in western Colorado and parts of Utah, southern Wyoming, northern New Mexico, and Arizona.

The UN Environment Program launches a campaign to save the world's great apes from extinction, asking for at least $1 million to get started.

The captain and crew of a tanker that spilled at least 185,000 gallons of diesel into the fragile marine environment of the Galapagos Islands have been arrested.

One hundred sixty-five countries approve the Kyoto rules aimed at halting global warming. The Kyoto Protocol requires industrial countries to scale back emissions of carbon dioxide and other greenhouse gases by an average of 5 percent from their 1990 levels by 2012. The United States, the world's biggest polluter rejects the pact.

The EPA reaches an agreement for the phaseout of a widely used pesticide, diazinon, because of potential health risks to children.

For the second time in three years, the average fuel economy of new passenger cars and light trucks sold in the United States dropped to its lowest level since 1980.

More and more Americans are breathing dirtier air, and larger U.S. cities such as Los Angeles and Atlanta remain among the worst for pollution.

In rural stretches of Alaska, global warming has thinned the Arctic pack ice, making travel dangerous for native hunters. Traces of industrial pollution from distant continents is showing up in the fat of Alaska's marine wildlife and in the breast milk of native mothers who eat a traditional diet including seal and walrus meat.

2002 A Congo volcano devastates a Congolese town burning everything in its path, creating a five-foot-high wall of cooling stone, and leaving a half million people homeless.

New research is conducted in the practice of killing sharks solely for their fins.

A report by the USGS shows the nation's waterways are awash in traces of chemicals used in beauty aids, medications, cleaners, and foods. Among the substances are caffeine, painkillers, insect repellent, perfumes, and nicotine. These substances largely escape regulation and defy municipal wastewater treatment.

A microbe is discovered to be a major cause of the destruction of beech trees in the northeastern United States.

A study discovers that, if fallen leaves are left in stagnant water, they can release toxic mercury, which eventually can accumulate in fish that live far downstream.

Scientists are experimenting with various sprays containing clay particles to kill toxic algae in seawater.

Meteorologists discover that the Mediterranean Sea receives air current pollutants from Europe, Asia, and North America.

Researchers report possible ways of blocking the deadly effects of anthrax.

2003 A new international treaty—The Protocol on Persistent Organic Pollutants (POPS) was ratified by 17 nations although the United States has not signed on. The treaty drafted by United Nations reduces and eliminates 16 toxic chemicals that are long-lived in the environment and travel globally. The new treaty, an extension of an earlier one signed in 2000, added four more organic persistent pollutants to the list.

Many global scientific studies reveal that excessive ultraviolet (UV) sunlight and pollution are linked to a decline in amphibian populations. Now Canadian biologists find that too much exposure of excessive UV radiation to tadpole populations reduces their chances of becoming frogs.

2003 marked the 50th anniversary of the research and publication of a different structure of the DNA model proposed by James D. Watson and

Francis H.C. Crick. In 1953 the scientists reported that the DNA molecule resembled a spiral staircase.

A new excavation in South Africa discovered the oldest fossils in the human family. The bones of a skull and a partial arm found in two caves date back to 4 million years ago according to scientists in Johannesburg

Scientists in New Jersey discovered that some outdoor antimosquito coils used to keep insects away can also cause respiratory health problems. The spiral-shaped container releases pollutants in the fumes expelled from coil. The researchers suggest that consumers should check these products carefully.

Researchers in Australia reported that pieces of plastic litter found in oceans continue to have an effect on marine wildlife. Small plastic chips are a hazard for seabirds who mistake the litter for food or fish eggs. The litter also moves up the food chain from fish that have ingested the plastic chips and in turn seals eat them.

2004 A scientific study reported that consumers should limit their consumption of farm-raised Atlantic salmon because of high concentrations of chlorinated organic contaminants in the fish. Their study revealed that the farm-raised salmon were contaminated with polychlorinated biphenyls (PCBs) and other organic chemicals. Except for the PCBs, the researchers agree that the farm-raised fish are healthy but consumption should be limited to no more than once a month in the diet. The researchers based their dietary report on the U.S. Environmental Protection Agency cancer risk assessments.

A group in Salisbury Plain, England is restoring Stonehenge to its natural setting. As a popular historic site to visitors, Stonehenge had become an area surrounded by roads and parking lots. The new restoration plan calls for building an underground tunnel for traffic and removing one of the roads. The present parking lots will become open grassy lawns.

Experts reported that two billion people lack reliable access to safe and nutritious food and 800 million, 40 percent of them children, are classified as chronically malnourished.

Public health officials in Uganda have reported progress in the country's fight against HIV, the AIDS virus. Since 1990's HIV cases in Uganda have dropped by more than 60 percent. Unfortunately, Uganda's neighboring countries are not doing well in their HIV prevention programs.

United Nations Secretary-General Kofi Annan stated, "by 2025, two-thirds of the world's population may be living in countries that face serious water shortages." The growing population is making surface water scarcer particularly in urban areas.

United States and Israel scientists have found a way to produce hydrogen from water. The hydrogen energy can be used in making fuel cells to power vehicles and homes. The research team uses solar radiation to heat sodium hydroxide in a solution of water. At high temperatures the water molecules (H_2O) break apart into oxygen and hydrogen. Using solar-power to produce hydrogen is better environmentally than hydrogen derived from fossil fuels.

APPENDIX B: ENDANGERED SPECIES BY STATE

The list below, obtained from the U.S. Fish and Wildlife Service, is an abridged listing of a selected group of endangered species (E) for each state. For a full list of endangered and threatened species, and other information about endangered species and the Endangered Species Act, see the Endangered Species Program Website at http://endangered.fws.gov/

ALABAMA

(Alabama has 106 plant and animal species that are listed as endangered (E) or threatened (T). The following list is only a selection of those plants and animals that are endangered. Contact the U.S. Fish and Wildlife Service to see the entire list.)

Animals

E - Bat, gray
E - Bat, Indiana
E - Cavefish, Alabama
T - Chub, spotfin
E - Clubshell, black
E - Combshell, southern
E - Darter, boulder
E - Fanshell
E - Kidneyshell, triangular
E - Lampmussel, Alabama
E - Manatee, West Indian
E - Moccasinshell, Coosa
E - Mouse, Alabama beach
E - Mussel, ring pink
E - Pearlymussel, cracking
E - Pearlymussel, Cumberland monkeyface
E - Pigtoe, dark
E - Plover, piping
E - Shrimp, Alabama cave
E - Snail, tulotoma (Alabama live-bearing)
E - Stork, wood
E - Turtle, Alabama redbelly (red-bellied)
E - Turtle, leatherback sea
E - Woodpecker, red-cockaded

Plants

E - Grass, Tennessee yellow-eyed
E - Leather-flower, Alabama
E - Morefield's leather-flower
E - Pinkroot, gentian
E - Pitcher-plant, Alabama canebrake
E - Pitcher-plant, green
E - Pondberry
E - Prairie-clover, leafy

ALASKA
Animals

E - Curlew, Eskimo (*Numenius borealis*)
E - Falcon, American peregrine (*Falco peregrinus anatum*)

Plant

E - Aleutian shield-fern (Aleutian holly-fern) (*Polystichum aleuticum*)

ARIZONA
Animals

E - Ambersnail, Kanab
E - Bat, lesser (Sanborn's) long-nosed
E - Bobwhite, masked (quail)
E - Chub, bonytail
E - Chub, humpback
E - Chub, Virgin River
E - Chub, Yaqui
E - Flycatcher, Southwestern willow
E - Jaguarundi
E - Ocelot
E - Pronghorn, Sonoran
E - Pupfish, desert
E - Rail, Yuma clapper
E - Squawfish, Colorado
E - Squirrel, Mount Graham red
E - Sucker, razorback
E - Topminnow, Gila (incl. Yaqui)
E - Trout, Gila
E - Vole, Hualapai Mexican
E - Woundfin

Plants

E – Arizona agave
E – Arizona cliffrose
E – Arizona hedgehog cactus
E – Brady pincushion cactus
E – Kearney's blue-star
E – Nichol's Turk's head cactus
E – Peebles Navajo cactus
E – Pima pineapple cactus
E – Sentry milk-vetch

ARKANSAS

Animals

E – Bat, gray
E – Bat, Indiana
E – Bat, Ozark big-eared
E – Beetle, American burying (giant carrion)
E – Crayfish, cave
E – Pearlymussel, Curtis'
E – Pearlymussel, pink mucket
E – Pocketbook, fat
E – Pocketbook, speckled
E – Rock-pocketbook, Ouachita (Wheeler's pearly mussel)
E – Sturgeon, pallid
E – Tern, least
E – Woodpecker, red-cockaded

Plants

E – Harperella
E – Pondberry
E – Running buffalo clover

CALIFORNIA

(California has more than 160 plant and animal species that are listed as endangered or threatened. The following list is only a selection of those plants and animals that are endangered. Contact the U.S. Fish and Wildlife Service to see the entire list.)

Animals

E – Butterfly, El Segundo blue
E – Butterfly, Lange's metalmark
E – Chub, Mohave tui
E – Condor, California
E – Crayfish, Shasta (placid)

E – Fairy shrimp, Conservancy
E – Falcon, American peregrine
E – Fly, Delhi Sands flower-loving
E – Flycatcher, Southwestern willow
E – Fox, San Joaquin kit
E – Goby, tidewater
E – Kangaroo rat, Fresno
E – Lizard, blunt-nosed leopard
E – Mountain beaver, Point Arena
E – Mouse, Pacific pocket
E – Pelican, brown
E – Pupfish, Owens
E – Rail, California clapper
E – Salamander, Santa Cruz long-toed
E – Shrike, San Clemente loggerhead
E – Shrimp, California freshwater
E – Snail, Morro shoulderband (banded dune)
E – Snake, San Francisco garter
E – Stickleback, unarmored threespine
E – Sucker, Lost River
E – Tadpole shrimp, vernal pool
E – Tern, California least
E – Toad, arroyo southwestern
E – Turtle, leatherback sea
E – Vireo, least Bell's
E – Vole, Amargosa

Plants

E – Antioch Dunes evening-primrose
E – Bakersfield cactus
E – Ben Lomond wallflower
E – Burke's goldfields
E – California jewelflower
E – California Orcutt grass
E – Clover lupine
E – Cushenbury buckwheat
E – Fountain thistle
E – Gambel's watercress
E – Kern mallow
E – Loch Lomond coyote-thistle
E – Robust spineflower (includes Scotts Valley spineflower)
E – San Clemente Island larkspur
E – San Diego button-celery
E – San Mateo thornmint
E – Santa Ana River woolly-star
E – Santa Barbara Island liveforever
E – Santa Cruz cypress
E – Solano grass
E – Sonoma sunshine (Baker's stickyseed)

E – Stebbins' morning-glory
E – Truckee barberry
E – Western lily

COLORADO
Animals

E – Butterfly, Uncompahgre fritillary
E – Chub, bonytail
E – Chub, humpback
E – Crane, whooping
E – Ferret, black-footed
E – Flycatcher, Southwestern willow
E – Plover, piping
E – Squawfish, Colorado
E – Sucker, razorback
E – Tern, least
E – Wolf, gray

Plants

E – Clay-loving wild-buckwheat
E – Knowlton cactus
E – Mancos milk-vetch
E – North Park phacelia
E – Osterhout milk-vetch
E – Penland beardtongue

CONNECTICUT
Animals

E – Mussel, dwarf wedge
E – Plover, piping
E – Tern, roseate
E – Turtle, hawksbill sea
E – Turtle, Kemp's (Atlantic) ridley sea
E – Turtle, leatherback sea

Plant

E – Sandplain gerardia

DELAWARE
Animals

E – Plover, piping
E – Squirrel, Delmarva Peninsula fox
E – Turtle, hawksbill sea
E – Turtle, Kemp's (Atlantic) ridley sea—Turtle,
      green sea
E – Turtle, leatherback sea

Plant

E – Canby's dropwort

FLORIDA

(Florida has more than 90 plant and animal species
that are listed as endangered or threatened. The following list is only a selection of those plants and
animals that are endangered. Contact the U.S. Fish
and Wildlife Service to see the entire list.)

Animals

E – Bat, gray
E – Butterfly, Schaus swallowtail
E – Crocodile, American
E – Darter, Okaloosa
E – Deer, key
E – Kite, Everglade snail
E – Manatee, West Indian (Florida)
E – Mouse, Anastasia Island beach
E – Mouse, Choctawahatchee beach
E – Panther, Florida
E – Plover, piping
E – Rabbit, Lower Keys
E – Rice rat (silver rice rat)
E – Sparrow, Cape Sable seaside
E – Stork, wood
E – Tern, roseate
E – Turtle, hawksbill sea
E – Turtle, Kemp's (Atlantic) ridley sea
E – Turtle, leatherback sea
E – Vole, Florida salt marsh
E – Woodpecker, red-cockaded
E – Woodrat, Key Largo

Plants

E – Apalachicola rosemary
E – Beautiful pawpaw
E – Brooksville (Robins') bellflower
E – Carter's mustard
E – Chapman rhododendron
E – Cooley's water-willow
E – Crenulate lead-plant
E – Etonia rosemary
E – Florida golden aster
E – Fragrant prickly-apple
E – Garrett's mint
E – Key tree-cactus
E – Lakela's mint
E – Okeechobee gourd

E – Scrub blazingstar
E – Small's milkpea
E – Snakeroot
E – Wireweed

GEORGIA
Animals

E – Acornshell, southern
E – Bat, gray
E – Bat, Indiana
E – Clubshell, ovate
E – Clubshell, southern
E – Combshell, upland
E – Darter, amber
E – Darter, Etowah
E – Kidneyshell, triangular
E – Logperch, Conasauga
E – Manatee, West Indian (Florida)
E – Moccasinshell, Coosa
E – Pigtoe, southern
E – Plover, piping
E – Stork, wood
E – Turtle, hawksbill sea
E – Turtle, Kemp's (Atlantic) ridley sea
E – Turtle, leatherback sea
E – Woodpecker, red-cockaded

Plants

E – American chaffseed
E – Black-spored quillwort
E – Canby's dropwort
E – Florida torreya
E – Fringed campion
E – Green pitcher-plant
E – Hairy rattleweed
E – Harperella
E – Large-flowered skullcap
E – Mat-forming quillwort
E – Michaux's sumac
E – Persistent trillium
E – Pondberry
E – Relict trillium
E – Smooth coneflower
E – Tennessee yellow-eyed grass

HAWAII

(Hawaii has 300 plant and animal species listed as endangered or threatened. The following list is only a selection of those plants and animals that are endangered. Contact the U.S. Fish and Wildlife Service to see the entire list.)

Animals

E – 'Akepa, Hawaii (honeycreeper)
E – Bat, Hawaiian hoary
E – Coot, Hawaiian
E – Creeper, Hawaiian
E – Crow, Hawaiian
E – Duck, Hawaiian
E – Duck, Laysan
E – Finch, Laysan (honeycreeper)
E – Finch, Nihoa (honeycreeper)
E – Goose, Hawaiian (nene)
E – Hawk, Hawaiian
E – Millerbird, Nihoa (old world warbler)
E – Nukupu'u (honeycreeper)
E – Palila (honeycreeper)
E – Parrotbill, Maui (honeycreeper)
E – Petrel, Hawaiian dark-rumped
E – Snails, Oahu tree
E – Stilt, Hawaiian
E – Turtle, hawksbill sea
E – Turtle, leatherback sea

Plants

E – Abutilon eremitopetalum
E – Bonamia menziesii
E – Carter's panicgrass
E – Diamond Head schiedea
E – Dwarf iliau
E – Fosberg's love grass
E – Hawaiian bluegrass
E – Hawaiian red-flowered geranium
E – Kaulu
E – Kiponapona
E – Mahoe
E – Mapele
E – Nanu
E – Nehe
E – Opuhe
E – Pamakani
E – Round-leaved chaff-flower
E – Viola helenae

IDAHO
Animals

E – Caribou, woodland
E – Crane, whooping

E – Limpet, Banbury Springs
E – Snail, Snake River physa
E – Snail, Utah valvata
E – Springsnail, Bruneau Hot
E – Springsnail, Idaho
E – Sturgeon, white
E – Wolf, gray

Plants

(No plants on the endangered list)

ILLINOIS

Animals

E – Bat, gray
E – Bat, Indiana
E – Butterfly, Karner blue
E – Dragonfly, Hine's emerald
E – Falcon, American peregrine
E – Fanshell
E – Pearlymussel, Higgins' eye
E – Pearlymussel, orange-foot pimple back
E – Pearlymussel, pink mucket
E,T – Plover, piping
E – Pocketbook, fat
E – Snail, Iowa Pleistocene
E – Sturgeon, pallid
E – Tern, least

Plant

E – Leafy prairie-clover

INDIANA

Animals

E – Bat, gray
E – Bat, Indiana
E – Butterfly, Karner blue
E – Butterfly, Mitchell's satyr
E – Clubshell
E – Fanshell
E – Mussel, ring pink (golf stick pearly)
E – Pearlymussel, cracking
E – Pearlymussel, orange-foot pimple back
E – Pearlymussel, pink mucket
E – Pearlymussel, tubercled-blossom
E – Pearlymussel, white cat's paw
E – Pearlymussel, white wartyback
E – Pigtoe, rough
E,T – Plover, piping

E – Pocketbook, fat
E – Riffleshell, northern
E – Tern, least

Plant

E – Running buffalo clover

IOWA

Animals

E – Bat, Indiana
E – Pearlymussel, Higgins' eye
E – Plover, piping
E – Snail, Iowa Pleistocene
E – Sturgeon, pallid
E – Tern, least

Plants

(No plants on the endangered list)

KANSAS

Animals

E – Bat, gray
E – Bat, Indiana
E – Crane, whooping
E – Curlew, Eskimo
E – Ferret, black-footed
E – Plover, piping
E – Sturgeon, pallid
E – Tern, least
E – Vireo, black-capped

Plants

(No plants on endangered list)

KENTUCKY

Animals

E – Bat, gray
E – Bat, Indiana
E – Bat, Virginia big-eared
E – Clubshell
E – Darter, relict
E – Falcon, American peregrine
E – Fanshell
E – Mussel, ring pink (golf stick pearly)
E – Mussel, winged mapleleaf
E – Pearlymussel, cracking
E – Pearlymussel, Cumberland bean

E - Pearlymussel, dromedary
E - Pearlymussel, little-wing
E - Pearlymussel, orange-foot pimple back
E - Pearlymussel, pink mucket
E - Pearlymussel, purple cat's paw
E - Pearlymussel, tubercled-blossom
E - Pearlymussel, white wartyback
E - Pigtoe, rough
E - Plover, piping
E - Pocketbook, fat
E - Riffleshell, northern
E - Riffleshell, tan
E - Shiner, Palezone
E - Shrimp, Kentucky cave
E - Sturgeon, pallid
E - Tern, least
E - Woodpecker, red-cockaded

Plants

E - Cumberland sandwort
E - Rock cress
E - Running buffalo clover
E - Short's goldenrod

LOUISIANA

Animals

E - Manatee, West Indian (Florida)
E - Pearlymussel, pink mucket
E - Pelican, brown
E - Plover, piping
E - Sturgeon, pallid
E - Tern, least
T - Turtle, green sea
E - Turtle, hawksbill sea
E - Turtle, Kemp's (Atlantic) ridley sea
E - Turtle, leatherback sea
E - Vireo, black-capped
E - Woodpecker, red-cockaded

Plants

E - American chaffseed
E - Louisiana quillwort
E - Pondberry

MAINE

Animals

E - Plover, piping
E - Tern, roseate
E - Turtle, leatherback sea

Plant

E - Furbish lousewort

MARYLAND

Animals

E - Bat, Indiana
E - Darter, Maryland
E - Mussel, dwarf wedge
E - Plover, piping
E - Squirrel, Delmarva Peninsula fox
E - Turtle, hawksbill sea
E - Turtle, Kemp's (Atlantic) ridley sea
E - Turtle, leatherback sea

Plants

E - Canby's dropwort
E - Harperella
E - Northeastern (Barbed bristle) bulrush
E - Sandplain gerardia

MASSACHUSETTS

Animals

E - Beetle, American burying (giant carrion)
E - Falcon, American peregrine
E - Mussel, dwarf wedge
E - Plover, piping
E - Tern, roseate
E - Turtle, hawksbill sea
E - Turtle, Kemp's (Atlantic) ridley sea
E - Turtle, leatherback sea
E - Turtle, Plymouth redbelly (red-bellied)

Plants

E - Northeastern (Barbed bristle)
E - Sandplain gerardia

MICHIGAN

Animals

E - Bat, Indiana
E - Beetle, American burying (giant carrion)
E - Beetle, Hungerford's crawling water
E - Butterfly, Karner blue
E - Butterfly, Mitchell's satyr
E - Clubshell
E - Plover, piping
E - Riffleshell, northern
E - Warbler, Kirtland's
E - Wolf, gray

Plant

E – Michigan monkey-flower

MINNESOTA

Animals

E – Butterfly, Karner blue
E – Mussel, winged mapleleaf
E – Pearlymussel, Higgins' eye
E – Plover, piping
E – Wolf, gray

Plant

E – Minnesota trout lily

MISSISSIPPI

Animals

E – Bat, Indiana
E – Clubshell, black (Curtus' mussel)
E – Clubshell, ovate
E – Clubshell, southern
E – Combshell, southern (penitent mussel)
E – Crane, Mississippi sandhill
E – Falcon, American peregrine
E – Manatee, West Indian (Florida)
E – Pelican, brown
E – Pigtoe, flat (Marshall's mussel)
E – Pigtoe, heavy (Judge Tait's mussel)
E – Plover, piping
E – Pocketbook, fat
E – Stirrupshell
E – Sturgeon, pallid
E – Tern, least
E – Turtle, hawksbill sea
E – Turtle, Kemp's (Atlantic) ridley sea
E – Turtle, leatherback sea
E – Woodpecker, red-cockaded

Plants

E – American chaffseed
E – Pondberry

MISSOURI

Animals

E – Bat, gray
E – Bat, Indiana
E – Bat, Ozark big-eared
E – Pearlymussel, Curtis'
E – Pearlymussel, Higgins' eye
E – Pearlymussel, pink mucket
E – Plover, piping
E – Pocketbook, fat
E – Sturgeon, pallid
E – Tern, least

Plants

E – Missouri bladderpod
E – Pondberry
E – Running buffalo clover

MONTANA

Animals

E – Crane, whooping
E – Curlew, Eskimo
E – Ferret, black-footed
E – Plover, piping
E – Sturgeon, pallid
E – Sturgeon, white
E – Tern, least
E – Wolf, gray

Plants

(No plants on endangered list)

NEBRASKA

Animals

E – Beetle, American burying
    (giant carrion)
E – Crane, whooping
E – Curlew, Eskimo
E – Ferret, black-footed
E – Plover, piping
E – Sturgeon, pallid
E – Tern, least

Plant

E – Blowout penstemon

NEVADA

Animals

E – Chub, bonytail
E – Chub, Pahranagat roundtail (bonytail)
E – Chub, Virgin River
E – Cui-ui
E – Dace, Ash Meadows speckled

E – Dace, Clover Valley speckled
E – Dace, Independence Valley speckled
E – Dace, Moapa
E – Poolfish (killifish), Pahrump
E – Pupfish, Ash Meadows Amargosa
E – Pupfish, Devils Hole
E – Pupfish, Warm Springs
E – Spinedace, White River
E – Springfish, Hiko White River
E – Springfish, White River
E – Sucker, razorback
E – Woundfin

Plants

E – Amargosa niterwort
E – Steamboat buckwheat

NEW HAMPSHIRE

Animals

E – Butterfly, Karner blue
E – Mussel, dwarf wedge
E – Turtle, leatherback sea

Plants

E – Jesup's milk-vetch
E – Northeastern (Barbed bristle) bulrush
E – Robbins' cinquefoil

NEW JERSEY

Animals

E – Bat, Indiana
E – Plover, piping
E – Tern, roseate
E – Turtle, hawksbill sea
E – Turtle, Kemp's (Atlantic) ridley sea
E – Turtle, leatherback sea

Plant

E – American chaffseed

NEW MEXICO

Animals

E – Bat, lesser (Sanborn's) long-nosed
E – Bat, Mexican long-nosed
E – Crane, whooping
E – Gambusia, Pecos

E – Isopod, Socorro
E – Minnow, Rio Grande silvery
E – Springsnail, Alamosa
E – Springsnail, Socorro
E – Sucker, razorback
E – Tern, least
E – Topminnow, Gila (incl. Yaqui)
E – Trout, Gila
E – Woundfin

Plants

E – Holy Ghost ipomopsis
E – Knowlton cactus
E – Kuenzler hedgehog cactus
E – Lloyd's hedgehog cactus
E – Mancos milk-vetch
E – Sacramento prickly-poppy
E – Sneed pincushion cactus
E – Todsen's pennyroyal

NEW YORK

Animals

E – Butterfly, Karner blue
E – Mussel, dwarf wedge
E, T – Plover, piping
E – Tern, roseate
E – Turtle, hawksbill sea
E – Turtle, Kemp's (Atlantic) ridley sea
E – Turtle, leatherback sea

Plants

E – Northeastern (Barbed bristle) bulrush
E – Sandplain gerardia

NORTH CAROLINA

Animals

E – Bat, Indiana
E – Bat, Virginia big-eared
E – Butterfly, Saint Francis' satyr
E – Elktoe, Appalachian
E – Falcon, American peregrine
E – Heelsplitter, Carolina
E – Manatee, West Indian (Florida)
E – Mussel, dwarf wedge
E – Pearlymussel, little-wing
E – Plover, piping
E – Shiner, Cape Fear
E – Spider, spruce-fir moss

E – Spinymussel, Tar River
E – Squirrel, Carolina northern flying
E – Tern, roseate
E – Turtle, hawksbill sea
E – Turtle, Kemp's (Atlantic) ridley sea
E – Turtle, leatherback sea
E – Wolf, red
E – Woodpecker, red-cockaded

Plants

E – American chaffseed
E – Bunched arrowhead
E – Canby's dropwort
E – Cooley's meadowrue
E – Green pitcher-plant
E – Harperella
E – Michaux's sumac
E – Mountain sweet pitcher-plant
E – Pondberry
E – Roan Mountain bluet
E – Rock gnome lichen
E – Rough-leaved loosestrife
E – Schweinitz's sunflower
E – Small-anthered bittercress
E – Smooth coneflower
E – Spreading avens
E – White irisette

NORTH DAKOTA

Animals

E – Crane, whooping
E – Curlew, Eskimo
E – Falcon, American peregrine
E – Ferret, black-footed
E – Plover, piping
E – Sturgeon, pallid
E – Tern, least
E – Wolf, gray

Plants

(No plants on endangered list)

OHIO

Animals

E – Bat, Indiana
E – Beetle, American burying (giant carrion)
E – Butterfly, Karner blue
E – Butterfly, Mitchell's satyr

E – Clubshell
E – Dragonfly, Hine's emerald
E – Fanshell
E – Madtom, Scioto
E – Pearlymussel, pink mucket
E – Pearlymussel, purple cat's paw
E – Pearlymussel, white cat's paw
E, T – Plover, piping
E – Riffleshell, northern

Plant

E – Running buffalo clover

OKLAHOMA

Animals

E – Bat, gray
E – Bat, Indiana
E – Bat, Ozark big-eared
E – Beetle, American burying
  (giant carrion)
E – Crane, whooping
E – Curlew, Eskimo
E – Plover, piping
E – Rock-pocketbook, Ouachita
E – Tern, least
E – Vireo, black-capped
E – Woodpecker, red-cockaded

Plants

(No plants on the endangered list)

OREGON

Animals

E – Chub, Borax Lake
E – Chub, Oregon
E – Deer, Columbian white-tailed
E – Pelican, brown
E – Sucker, Lost River
E – Sucker, shortnose
E – Turtle, leatherback sea

Plants

E – Applegate's milk-vetch
E – Bradshaw's desert-parsley
E – Malheur wire-lettuce
E – Marsh sandwort
E – Western lily

PENNSYLVANIA

Animals

E – Bat, Indiana
E – Clubshell
E – Mussel, dwarf wedge
E – Mussel, ring pink (golf stick pearly)
E – Pearlymussel, cracking
E – Pearlymussel, orange-foot pimple back
E – Pearlymussel, pink mucket
E – Pigtoe, rough
E, T – Plover, piping
E – Riffleshell, northern

Plant

E – Northeastern (Barbed bristle)
   bulrush

RHODE ISLAND

Animals

E – Beetle, American burying
E – Falcon, American peregrine
E – Plover, piping
E – Tern, roseate
E – Turtle, hawksbill sea
E – Turtle, Kemp's
E – Turtle, leatherback sea

Plant

E – Sandplain gerardia

SOUTH CAROLINA

Animals

E – Bat, Indiana
E – Heelsplitter, Carolina
E – Manatee, West Indian (Florida)
E – Plover, piping
E – Stork, wood
E – Tern, roseate
E – Turtle, hawksbill sea
E – Turtle, Kemp's (Atlantic) ridley sea
E – Turtle, leatherback sea
E – Woodpecker, red-cockaded

Plants

E – American chaffseed
E – Black-spored quillwort

E – Bunched arrowhead
E – Canby's dropwort
T – Dwarf-flowered heartleaf
E – Harperella
E – Michaux's sumac
E – Mountain sweet pitcher-plant
E – Persistent trillium
E – Pondberry
E – Relict trillium
E – Rough-leaved loosestrife
E – Schweinitz's sunflower
E – Smooth coneflower

SOUTH DAKOTA

Animals

E – Beetle, American burying
   (giant carrion)
E – Crane, whooping
E – Curlew, Eskimo
E – Ferret, black-footed
E – Plover, piping
E – Sturgeon, pallid
E – Tern, least
E – Wolf, gray

Plants

(No plants on the endangered list)

TENNESSEE

(Tennessee has 81 plant and animal species that are listed as endangered or threatened. The following list is only a selection of those plants and animals that are endangered. Contact the U.S. Fish and Wildlife Service to see the entire list.)

Animals

E – Bat, gray
E – Bat, Indiana
E – Combshell, upland
E – Crayfish, Nashville
E – Darter, amber
E – Fanshell
E – Lampmussel, Alabama
E – Madtom, Smoky
E – Marstonia (snail), (royalobese)
E – Moccasinshell, Coosa
E – Mussel, ring pink (golf stick pearly)
E – Pearlymussel, Appalachian monkeyface

E - Pearlymussel, Cumberland bean
E - Riversnail, Anthony's
E - Spider, spruce-fir moss
E - Squirrel, Carolina northern flying
E - Sturgeon, pallid
E - Tern, least
E - Wolf, red
E - Woodpecker, red-cockaded

Plants

E - Cumberland sandwort
E - Green pitcher-plant
E - Large-flowered skullcap
E - Leafy prairie-clover (Dalea)
E - Roan Mountain bluet
E - Rock cress
E - Rock gnome lichen
E - Ruth's golden aster
E - Spring Creek bladderpod
E - Tennessee purple coneflower
E - Tennessee yellow-eyed grass

TEXAS

(Texas has 70 plant and animal species that are listed as endangered or threatened. The following list is only a selection of those plants and animals that are endangered. Contact the U.S. Fish and Wildlife Service to see the entire list.)

Animals

E - Bat, Mexican long-nosed
E - Beetle, Coffin Cave mold
E - Crane, whooping
E - Curlew, Eskimo
E - Darter, fountain
E - Falcon, northern aplomado
E - Jaguarundi
E - Manatee, West Indian (Florida)
E - Minnow, Rio Grande silvery
E - Ocelot
E - Pelican, brown
E - Plover, piping
E - Prairie-chicken, Attwater's greater
E - Pupfish, Comanche Springs
E - Salamander, Texas blind
E - Spider, Tooth Cave
E - Tern, least
E - Toad, Houston
E - Turtle, hawksbill sea

E - Turtle, Kemp's (Atlantic) ridley sea
E - Vireo, black-capped
E - Warbler, golden-cheeked
E - Woodpecker, red-cockaded

Plants

E - Ashy dogweed
E - Black lace cactus
T - Hinckley's oak
E - Large-fruited sand-verbena
E - Little Aguja pondweed
E - Lloyd's hedgehog cactus
E - Nellie cory cactus
E - Sneed pincushion cactus
E - South Texas ambrosia
E - Star cactus
E - Terlingua Creek cats-eye
E - Texas poppy-mallow
E - Texas snowbells
E - Texas wild-rice
E - Tobusch fishhook cactus
E - Walker's manioc

UTAH

Animals

E - Ambersnail, Kanab
E - Chub, bonytail
E - Chub, humpback
E - Chub, Virgin River
E - Crane, whooping
E - Ferret, black-footed
E - Flycatcher, Southwestern willow
E - Snail, Utah valvata
E - Squawfish, Colorado
E - Sucker, June
E - Sucker, razorback
E - Woundfin

Plants

E - Autumn buttercup
E - Barneby reed-mustard
E - Barneby ridge-cress (peppercress)
E - Clay phacelia
E - Dwarf bear-poppy
E - Kodachrome bladderpod
E - San Rafael cactus
E - Shrubby reed-mustard (toad-flax cress)
E - Wright fishhook cactus

VERMONT

Animals

E – Bat, Indiana
E – Mussel, dwarf wedge

Plants

E – Jesup's milk-vetch
E – Northeastern (Barbed bristle) bulrush

VIRGINIA

Animals

E – Bat, gray
E – Bat, Indiana
E – Bat, Virginia big-eared
E – Darter, duskytail
E – Falcon, American peregrine
E – Fanshell
E – Isopod, Lee County cave
E – Logperch, Roanoke
E – Mussel, dwarf wedge
E – Pearlymussel, Appalachian monkeyface
E – Pearlymussel, birdwing
E – Pearlymussel, cracking
E – Pearlymussel, Cumberland monkeyface
E – Pearlymussel, dromedary
E – Pearlymussel, green-blossom
E – Pearlymussel, little-wing
E – Pearlymussel, pink mucket
E – Pigtoe, fine-rayed
E – Pigtoe, rough
E – Pigtoe, shiny
E – Plover, piping
E – Riffleshell, tan
E – Salamander, Shenandoah
E – Snail, Virginia fringed mountain
E – Spinymussel, James River (Virginia)
E – Squirrel, Delmarva Peninsula fox
E – Squirrel, Virginia northern flying
E – Turtle, hawksbill sea
E – Turtle, Kemp's (Atlantic) ridley sea
E – Turtle, leatherback sea
E – Woodpecker, red-cockaded

Plants

E – Northeastern (Barbed bristle) bulrush
E – Peter's Mountain mallow
E – Shale barren rock-cress
E – Smooth coneflower

WASHINGTON

Animals

E – Caribou, woodland
E – Deer, Columbian white-tailed
E – Pelican, brown
E – Turtle, leatherback sea
E – Wolf, gray

Plants

E – Bradshaw's desert-parsley
  (lomatium)
E – Marsh sandwort

WEST VIRGINIA

Animals

E – Bat, Indiana
E – Bat, Virginia big-eared
E – Clubshell
E – Falcon, American peregrine
E – Fanshell
E – Mussel, ring pink
E – Pearlymussel, pink mucket
E – Pearlymussel, tubercled-blossom
E – Riffleshell, northern
E – Spinymussel, James River
E – Squirrel, Virginia northern flying

Plants

E – Harperella
E – Northeastern (Barbed bristle)
  bulrush
E – Running buffalo clover
E – Shale barren rock-cress

WISCONSIN

Animals

E – Butterfly, Karner blue
E – Dragonfly, Hine's emerald
E – Mussel, winged mapleleaf
E – Pearlymussel, Higgins' eye
E, T – Plover, piping
E – Warbler, Kirtland's
E – Wolf, gray

Plants

(No plants on endangered list)

WYOMING

Animals

E – Crane, whooping
E – Dace, Kendall Warm Springs
E – Ferret, black-footed
E – Squawfish, Colorado

E – Sucker, razorback
E – Toad, Wyoming
E – Wolf, gray

Plants

(No plants on endangered list)

APPENDIX C: WEBSITES BY CLASSIFICATION

Please note that the authors have made a consistent effort to include up-to-date Websites. However, over time, some Websites may move or no longer be posted.

ACID MINE DRAINAGE

National Reclamation Center, West Virginia University, Evansdale office, http//www.nrcce.wvu.edu/.

ACID RAIN

http://www.epa.gov/docs/acidrain/andhome/html.

The EPA has a hotline to request educational materials or respond to questions regarding acid rain: (202) 343–9620. http://www.econet.apc.org/acid rain.

Environmental Protection Agency, http://www.epa.gov/docs/acidrain/effects/enveffct.html.

National Reclamation Center's West Virginia University, Evansdale office: http://www.nrcce.wvu.edu/

USGS Water Science/Acid Rain, http://wwwga.usgs.gov/edu/acidrain.html.

AGENCY FOR TOXIC SUBSTANCES AND DISEASES

Registry Division of Toxicology
1600 Clifton Road NE Mailstop E-29
Atlanta, GA 30333
Website: http://www.atsdr1.atsdr.cdc.gov:8080/atsdrhome.html.

Agency for Toxic Substances and Disease Registry, http://www.atsdr.cdc.gov/cxcx3.html.

Information on biosphere reserves and UNESCO's Man and the Biosphere Programme, UNESCO: http://www.unesco. org

Man and the Biosphere Program: http://www. mabnet.org

AGRICULTURE

United States Department of Agriculture, http://www.usda.gov.

ALTERNATIVE FUELS

Department of Energy, http://www.doe.gov.

Department of Energy Alternative Fuels Data Center, http://www.afdc.nrel.gov; http://www.afdc.doe.gov/; or http://www.fleets.doe.gov.

AMPHIBIANS

http://www.frogweb.gov/

ANTARCTICA

Antarctica Treaty, http://www.sedac.ciesin.org/pidb/register/reg-024.rrr.html.

Greenpeace International Antarctic Homepage, http://www.greenpeace.org/~comms/98/antarctic.

International Centre for Antarctic Information and Research Homepage (includes text of Antarctic Treaty), http://www.icair.iac.org.nz.

Virtual Antarctica, http://www.exploratorium.edu

ARCTIC

Arctic Circle (University of Connecticut), http://arcticcircle.uconn.edu/arcticcircle.

Arctic Council Home Page, http://www.nrc.ca/arctic/index.html.

Arctic Monitoring and Assessment Programme (Norway), http://www.gsf.de/ UNEP/amap1.html.

Arctic National Wildlife Refuge, http://energy.usgs.gov/factsheets/ANWR/ANWR.html.

Institute of Arctic and Alpine Research, http://instaar.colorado.edu.

Institute of the North (Alaska Pacific University),

Inuit Circumpolar Conference,
NOAA Fisheries, http://www.nmfs.gov/.

Nunavut,
Smithsonian Institution Arctic Studies Center,
http://www.mnh.si.edu/arctic.

U.S. Fish and Wildlife Service
U.S. Department of the Interior

1849 C Street, NW,
Washington, D.C. 20240
Telephone: (202) 208-5634
Website: http://www.fws.gov.

World Conservation Monitoring Centre Arctic
Programme, http://www.wcmc.org.uk/
arctic.

AUTOMOBILE

Cars and Their Enviromental Impact,
http://www.environment.volvocars.com/
ch1-1.htm.

National Center for Vehicle Emissions Control
and Safety (NCVECS), http://www.colostate.
edu/Depts/NCVECS/ncvecs1.html.

U.S. Environmental Protection Agency Fact Sheet
(EPA 400-F-92-004, August 1994), "Air Toxics
from Motor Vehicles," http://www.epa.gov/
oms/02-toxic.htm.

U.S. Enviromental Protection Agency, Office of
Mobile Sources, http://www.epa.gov/oms.

BIOLOGICAL WEAPONS

Federation of American Scientist Biological
Weapons Control, http://www.fas.org/bwc.

Chemical and Biological Defense Information
Analysis Center, http://www.cbiac.apgea.
army. mil

BIOMES

Committee for the National Institute for the
Environment, http://www.cnie.org/nle/
biodv-6.html.

BIOREMEDIATION

Consortium, http://www.rtdf.org/public/
biorem.

BROWNFIELD

Projects, http://www.epa.gov/brownfields/.

CERES

Website: http://www.ceres.org or
e-mail ceres@igc.apc.org.
Summaries of Major Environmental Laws,
http://www.epa.gov/region5/defs/index.html.

CHEETAHS

Cheetah Conservation Fund

4649 Sunnyside Avenue N, Suite 325
Seattle, WA 98103
Website: http://www.cheetah.org.

World Wildlife Fund

1250 24th Street, NW,
Washington, D.C. 20037
Telephone: 1-800-225-5993
Website: http://www.worldwildlife.org/.

CHEMICAL WEAPONS

Chemical Stockpile Disposal Project (CSDP),
http://www.pmcd.apgea.army.mil/
graphical/CSDP/index.html.

Tooele Chemical Agent Disposal Site Facility,
http://www.deq.state.ut.us/eqshw/cds/
tocdfhp1.htm.

CLEAN WATER ACT

Sierra Club, "Happy 25th Birthday, Clean
Water Act," http://sierraclub.org/wetlands/
cwabday.html.

CLIMATE CHANGE AND GLOBAL WARMING

U.S. Geological Survey, Climate Change and
History, http://geology.usgs.gov/index.shtml.

EPA Global Warming Site,
http://www.epa.gov/globalwarming.

Greenpeace International, Climate,
http://www.greenpeace.org/~climate.

United Nations Intergovernmental Panel on
Climate Change, http://www.ipcc.ch.

COAL

Coal Age Magazine, http://coalage.com.

Department of Energy, Office of Fossil Energy, http:/www.doe.gov.

U.S. Geological Survey, National Coal Resources Data System, http:energy.er.usgs.gov/ coalqual. htm.

COASTAL AND MARINE GEOLOGY

U.S. Geological Survey, http://marine.usgs.gov/.

COMPOSTING

EPA Office of Solid Waste and Emergency Response—Composting, http:www.epa. gov/epaoswer/non-hw/compost/ index.htm

Cornell Composting, http://www.cfe.cornell. edu/compost/Composting_Homepage.html

CONSENT DECREES

EPA Office of Enforcement and Compliance Assurance, http://es.epa.gov/oeca/osre/ decree.html.

CORAL REEFS

Coral Reef Alliance, http://www.coral.org.

Coral Reef Network Directory, Greenpeace
1436 U Street, NW
Washington, D.C. 20009
Website: http://www.greenpeace.org.

EARTHDAY 2000

Earth Day Network

91 Marion Street,
Seattle, WA 98104
Telephone: 1(206)-264-0114.
Website: http://www.earthday.net/;
and worldwide@earthday.net.

EARTHWATCH

Earthwatch Institute International, http://www. earthwatch.org.

EL NIÑO

El Niño/La Niña theme page, contact NOAA
Website: http://www.pmel.noaa.gov/toga-tao/ el-nino/nino-home-low.html.

NOAA, La Niña homepage, www.elnino.noaa. gov/lanina.html.

National Center for Atmospheric Research, http://www.ncar.ucar.edu/.

National Hurricane Center/Tropical Prediction Center, http://www.nhc.noaa.gov/.

National Oceanographic and Atmospheric Administration, http://www.noaa.gov/.

Scripps Institute of Oceanography, http://sio.ucsd.edu/supp_groups/siocomm/ elnino/elnino.html.

ELECTRIC VEHICLES

Electric Vehicle Association of the Americas 800-438-3228, http://www.evaa.org.

Electric Vehicle Technology, http://www.avere.org/.

ELEPHANTS

African Wildlife Foundation, http://www.awf.org.

U.S. Fish and Wildlife Service, Species List of Endangered and Threatened Wildlife, http://endangered.fws.gov/

World Wildlife Fund, http://www.wwf.org.

ETHANOL

U.S. Department of Energy, Energy Efficiency and Renewable Energy Clearinghouse,

P.O. Box 3048
Merrifield, VA 22116
E-mail: energyinfo@delphi.com.
Website: http://www.doe.gov.

EVERGLADES

National Park Service, Everglades National Park, http://www.nps.gov/ever.

FEDERAL EMERGENCY MANAGEMENT AGENCY (FEMA)

FEMA, http://www.fema.gov.

FISHING, COMMERCIAL

National Oceanographic and Atmospheric Administration Fisheries, http://www.nmfs. gov/.

United Nations Food and Agriculture Organization Fisheries, http://www.fao.org/waicent/faoinfo/fishery/fishery.htm.

FORESTS

American Forests, http://www.amfor.org.

Greenpeace International, Forests, http://www. greenpeace.org/~forests.

Society of American Foresters, http://www. safnet.org.

U.S. Forest Service, http://www.fs.fed.us.

U.S. Forest Service Research, http://www.fs.fed.us/links/research.shtml.

World Conservation Monitoring Centre, http://www.wcmc.org.uk.

World Resources Institute Forest Frontiers Initiative, http://www.wri.org/ffi.

World Wildlife Fund (Worldwide Fund for Nature) Forests for Life Campaign, http://www.panda.org/forests4life.

FUEL CELLS AND OTHER ALTERNATIVE FUELS

Crest's Guide to the Internet's Alternative Energy Resources, http://solstice.crest.org/online/aeguide/aehome.html.

U.S. Department of Energy

P.O. Box 12316
Arlington, VA 22209
Telephone: 1–800–423–1363
Website: http://www.doe.gov.

U.S. Department of Energy, Alternative Fuels Data Center, http://www.afdc.nrel.gov.

GEOLOGY

Geological surveys, U.S. Geological Survey, http://www.usgs.gov/.

For general interest publications and products, http://mapping.usgs.gov/www/products/mappubs.html.

GEOTHERMAL SITES

Energy and Geoscience Institute

University of Utah
423 Wakara Way

Salt Lake City, UT 84108
Website: http://www.egi.utah.edu.

Geothermal energy information, http://geothermal.marin.org.

Geothermal database USA and Worldwide, http://www.geothermal.org.

International geothermal, http://www.demon.co.uk/geosci/igahome.html.

Solstice is the Internet information service of the Center for Renewable Energy and Sustainable Technology (CREST), http://solstice.crest. org/

GLACIERS SHRINKING

United States Geological Survey, Climate Change and History, http://geology.usgs.gov/index. shtml.

Sierra Club, Public Information Center, (415) 923-5653; or the Global Warming and Energy Team, (202) 547-1141, or by E-mail: information@sierraclub.org.

GLOBEC

Educational Website, http://cbl.umces.edu/fogarty/usglobec/misc/education.html.

GRASSLANDS AND PRAIRIES

Postcards from the Prairie, http://www.nrwrc.usgs.gov/postcards/postcards.htm.

University of California, Berkeley, World Biomes, Grasslands, http://www.ucmp.berkeley.edu/glossary/gloss5/biome/grasslan.html.

Worldwide Fund for Nature, Grasslands and Its Animals, http://www.panda.org/kids/wildlife/idxgrsmn.htm.

GROUNDWATER

EPA, http://www.epa.gov/swerosps/ej/.

Groundwater atlas of the United States, http://www.capp.er.usgs.gov/publicdocs/gwa/.

HAZARDOUS MATERIALS TRANSPORTAION ACT

Website: http://www.dot.gov.

HAZARDOUS SUBSTANCES

U.S. Environmental Protection Agency Program, http://epa.gov/.

U.S. Occupational Safety and Health Administration (OSHA), http://www.osha.gov/toxicsubstances/index.html.

Environmental Defense Fund (data on wastes and chemicals at U.S. sources), http://www.scorecard.org.

HAZARDOUS WASTE TREATMENT

Federal Remedial Technologies Roundtable, Hazardous Waste Clean-Up Information ("CLU-IN"), http://www.clu-in.org.

HEAVY METALS

U.S. Environmental Protection Agency, Office of Pollution Prevention and Toxics, http://www.epa.gov/opptintr.

HIGH-LEVEL RADIOACTIVE WASTES

U.S. Nuclear Regulatory Commission, Radioactive Waste Page, http://www.nrc.gov/NRC/radwaste.

U.S. Environmental Protection Agency, Mixed-Waste Homepage, http://www.epa.gov/radiation/mixed-waste.

HURRICANES

National Hurricane Center, http://www.nhc.noaa.gov.

HYDROELECTRIC POWER

U.S. Bureau of Reclamation Hydropower Information, http://www.usbr.gov/power/edu/edu.htm.

U.S. Geological Survey, http://wwwga.usgs.gov/edu/hybiggest.html.

HYDROGEN

National Renewable Energy Laboratory, http://www.nrel.gov/lab/pao/hydrogen.html.

EnviroSource, Hydrogen InfoNet, http:///www.eren.doe.gov/hydrogen/infonet.html.

INTERNATIONAL ATOMIC ENERGY AGENCY

Agency, http://www.iaea.org.

Managing Radioactive Waste Fact Sheet, http://www.iaea.org/worldatom/inforesource/factsheets/manradwa.html.

INTERNATIONAL COUNCIL FOR LOCAL ENVIRONMENTAL INITIATIVES

Homepage, http://www.iclei.org.

INTERNATIONAL REGISTER OF POTENTIALLY TOXIC CHEMICALS

Homepage, http://www.unep.org/unep/program/hhwb/chemical/irptc/home.htm.

INTERNATIONAL WHALING COMMISSION

Homepage, http://www.ourworld.compuserve.com/homepages/iwcoffice.

INVERTEBRATES: THREATENED AND ENDANGERED

U.S. Fish and Wildlife Service, Species List of Endangered and Threatened Wildlife, http://endangered.fws.gov/

LANDSAT AND SATELLITE IMAGES

Earthshots, Satellite Images of Environmental Change, http://www.usgs.gov/Earthshots/.

Landsat Gateway, http://landsat.gsfc.nasa.gov/main.htm.

LEAD

National Lead Information Center's Clearinghouse, 1-800-424-LEAD, http://www.epa.gov/lead/.

LEOPARDS

U.S. Fish and Wildlife Service, Species List of Endangered and Threatened Wildlife, http://www.fws.gov/r9endspp/lsppinfo.html.

LITTER

Keep America Beautiful, http://www.kab.org.

MAMMALS

U.S. Fish and Wildlife Service, Vertebrate Animals, http://www.fws.gov/r9endspp/lsppinfo.html.

MANATEES

Save the Manatees, http://www.savethemanatee. org.

Sea World, Manatees, http://www.seaworld.org/manatee/sciclassman.html.

MARSHES

Environmental Protection Agency, Office of Wetlands, Oceans, Watersheds, http://www.epa.gov/owow/wetlands/wetland2.html.

North American Waterfowl and Wetlands Office, http://www.fws.gov/r9nawwo.

North American Wetlands Conservation Act, http://www.fws.gov/r9nawwo/nawcahp.html.

North American Wetlands Conservation Council, http://www.fws.gov/r9nawwo/nawcc.html.

Wetlands, wetlands-hotline@epamail.epa.gov.

MATERIAL SAFETY DATA SHEET

Toxic chemicals, http://www.siri.org/msds; http://www.ilpi.com/mads/index.html.

MENDES, CHICO

Chico Mendes, http://www.edf.org/chico.

NATURAL DISASTERS

Building Safer Structures, http://quake.wr.usgs. gov/QUAKES/FactSheets/SaferStructures/.

Center for Integration of Natural Disaster Information, http://cindi.usgs.gov/events/.

Earthquakes, http://quake.wr.usgs.gov/; http://geology.usgs.gov/quake.html. For the latest earthquake information http://quake.wr.usgs.gov/QUAKES/CURRENT/current.html

National Hurricane Center, http://www.nhc.noaa.gov.

U.S. Geological Survey, http://geology.usgs.gov/whatsnew.html.

NATIONAL MARINE FISHERIES

History of National Marine Fisheries Service, http://www.wh.whoi.edu/125th/history/century.html.

National Marine Fisheries, http://kingfish.ssp.nmfs.gov.

NOAA Fisheries, http://www.nmfs.gov/.

NATIONAL OCEAN AND ATMOSPHERIC ADMINISTRATION (NOAA)

Climate forecasting, http://www.cdc.noaa.gov/ Seasonal/.

El Niño Theme Page, http://www.pmel.noaa.gov/toga-tao/el-nino/nino-home-low.html.

Homepage, http://www.noaa.gov/.

Recover Protected Species, http://www.noaa.gov/nmfs/recover.html.

Safe Navigation Page, http://anchor.ncd.noaa.gov/psn/psn.htm.

NATIONAL WEATHER SERVICE

Homepage, http://www.nws.noaa.gov.

NATIONAL WILDLIFE REFUGE SYSTEM

Homepage, http://refuges.fws.gov/NWRSHomePage.html.

NATURAL GAS

American Gas Association, http://www.aga.org.

Oil and Gas Journal Online, http://www.ogjonline.com.

U.S. Department of Energy, Energy Information Administration, http://www.eia.doe.gov.

U.S. Department of Energy, Office of Fossil Energy, http://www.fe.doe.gov.

U.S. Geological Survey Energy, Resources Program, http://energy.usgs.gov/index.html.

NOISE POLLUTION

Noise Pollution Clearinghouse, http://www. nonoise.org.

NONPOINT SOURCES

Nonpoint Source Pollution Control Program, http://www.epa.gov/OWOW/NPS/ whatudo.html; http://www.epa.gov/ OWOW/ NPS/.

NUCLEAR ENERGY AND NUCLEAR REACTORS

American Nuclear Society, http://www.ans.org.

Nuclear Energy Institute, http://www.nei.org.

Nuclear Information and Resource Service, http://www.nirs.org.

U.S. Department of Energy, Office of Nuclear Energy, Science and Technology, http://www.ne.doe.gov.

U.S. Nuclear Regulatory Commission, http://www.nrc.gov.

NUCLEAR WASTE POLICY ACT

American Nuclear Society, http://www.ans.org.

Nuclear Energy Institute, http://www.nei.org.

NUCLEAR WASTE SITES

Hazard Ranking System, http://www.epa. gov/ superfund/programs/npl_hrs/ hrsint.htm.

National Research Council, Board on Radioactive Waste Management, http://www4.nas.edu/ brwm/brwm-res.nsf.

Superfund, http://www.pin.org/superguide.htm; http://www.epa.gov/superfund.

U.S. Department of Energy, Office of Civilian Radioactive Waste Management, http://www.rw.doe.gov.

U.S. Environmental Protection Agency, Mixed-Waste Homepage, http://www.epa. gov/radiation/mixed-waste.

U.S. Nuclear Regulatory Commission, Radioactive Waste Page, http://www.nrc.gov/ NRC/ radwaste.

OCCUPATIONAL SAFETY AND HEALTH ACT (OSHA)

OSHA Homepage, http://www.osha.gov.

OCEAN THERMAL ENERGY CONVERSION (OTEC)

National Renewable Energy Laboratory

1617 Cole Boulevard
Golden, CO 80401
Website: http:llnrelinfo.nrel.gov.

Natural Energy Laboratory of Hawaii, http://bigisland.com/nelha/index.html.

OCEANS

National Oceanographic and Atmospheric Administration, http://www.noaa.gov/.

Safe Ocean Navigation Page, http://anchor.ncd. noaa.gov/psn/psn.htm.

OFFICE OF SURFACE MINING

Office of Surface Mining, http://www.osmre.gov.

Appalachian Clean Streams Initiative, majordomo@osmre.gov.

OLD-GROWTH FORESTS

Greenpeace International, Forests, http://www.greenpeace.org/~forests.

World Resources Institute, Forest Frontiers Initiative, http://www.wri.org/ffi.

OLMSTEAD, FREDERICK LAW

Homepage, http://fredericklawolmsted.com.

ORGANIZATION OF PETROLEUM EXPORTING COUNTRIES (OPRC)

Homepage, http://www.opec.org.

OVERFISHING

Information and data statistics, http://www.nmfs. gov.

National Aeronautics and Space Administration, Ocean Planet, http://seawifs.gsfc.nasa.gov/ OCEAN_PLANET/HTML/ peril_overfishing.html.

National Marine Fisheries Service, http://www. nmfs.gov.

NOAA, http://www.noaa.gov.

United Nations Food and Agricultural Organization, http://www.fao.org.

United Nations Food and Agriculture Organization Fisheries, http://www.fao.org/.

United Nations System, http://www.unsystem.org.

OZONE-RELATED ISSUES

Environmental Protection Agency, science of ozone depletion, http://www.epa.gov/ozone/science/.

NOAA, Commonly Asked Questions about Ozone, www.publicaffairs.noaa.gov/grounders/ozo1.html.

NOAA, Network for the Detection of Stratospheric Change, www.noaa.gov.

PARROTS

Online Book of Parrots, http://www.ub.tu-clausthal.dep/p_welcome.html.

World Parrot Trust, http://www.worldparrottrust.org.

World Wildlife Fund, http:www.panda.org.

PESTICIDES

Toxics and Pesticides, http://www.epa.gov/oppfead1/work_saf/.

Pesticides in the Atmosphere, http://ca.water.usgs.gov/pnsp/atmos.

PETERSON, ROGER TORY

Roger Tory Peterson Institute of Natural History,

311 Curtis Street
Jamestown, NY 14701
Website: http://www.rtpi.org/info/rtp.htm.

PETROLEUM

American Petroleum Institute, http://www.api.org.

Petroleum Information, http://www.petroleuminformation.com.

Oil and Gas Journal Online, http://www. ogjonline.com.

U.S. Department of Energy, Energy Information Administration, http://www.eia.doe.gov.

U.S. Department of Energy, Office of Fossil Energy, http://www.fe.doe.gov.

U.S. Geological Survey Energy Resources Program, http://energy.usgs.gov/index.html.

U.S. Geological Survey Fact Sheet FS-145-97, "Changing Perceptions of World Oil and Gas Resources as Shown by Recent USGS Petroleum Assessments," http://greenwood.cr.usgs.gov/pub/fact-sheets/fs-0145-97/fs-0145-97.html.

PLUTONIUM

U.S. Nuclear Regulatory Commission, Radioactive Waste Page, http://www.nrc.gov/NRC/radwaste.

RADIATION AND RADIOACTIVE WASTES

International Atomic Energy Agency, "Managing Radioactive Waste" Fact Sheet, http://www.iaea.org/worldatom/inforesource/factsheets/manradwa.html.

National Research Council, Board on Radioactive Waste Management, http://www4.nas.edu/brwm/brwm-res.nsf.

U.S. Department of Energy, Office of Civilian Radioactive Waste Management, http://www. rw.doe.gov.

U.S. Environmental Protection Agency, Mixed-Waste Homepage, http://www.epa.gov/radiation/mixed-waste.

U.S. Nuclear Regulatory Commission, Radioactive Waste Page, http://www.nrc.gov/NRC/radwaste.

RADON

Radon in Earth, Air, and Water, http://sedwww.cr.usgs.gov:8080/radon/radonhome.html.

RAIN FORESTS

Greenpeace International, forests, http://www.greenpeace.org/~forests.

Rainforest Action Network (RAN)

President Randy Hayes
221 Pine Street Suite 500
San Francisco, CA 94104
Telephone: (415) 398-4404
Website: http://www.ran.org

Rainforest Alliance (RA)

65 Bleeker Street
New York, NY 10012
Website: http://www.rainforest-alliance.org

U.S. Forest Service, http://www.fs.fed.us.

World Wildlife Fund (Worldwide Fund for Nature), Forests for Life Campaign, http://www.panda.org/forests4life.

RESOURCE CONSERVATION AND RECOVERY ACT

Homepage, http://www.epa.gov/epaoswer/hotline.

SALMON

National Marine Fisheries Service, http://www.nwr.noaa.gov/1salmon/salmesa/index.htm.
NOAA Fisheries, http://www.nmfs.gov/.

SALT MARSHES

National Wetlands Research Center, http://www.nwrc.usgs.gov/educ_out.html.

USGS Coastal and Marine Geology, http://marine.usgs.gov/.

SANITARY LANDFILLS

Solid waste management, http://web.mit.edu/urbanupgrading/urban environment/

Landfills - Solid and Hazardous Waste and Ground-water Quality Protection, http://www.gfredlee.com/plandfil2.htm

SIBERIA

Siberia, http://www.cnit.nsk.su/univer/english/siberia.htm.

SOLAR ENERGY

American Solar Energy Society

2400 Central Avenue, Suite G-1
Boulder, CO 80301.
Website: http://www.soton.ac.uk/~solar/.

Solar Energy Industries Association

122 C Street, NW, 4th Floor
Washington, D.C. 20001.
Website: http://www.seia.org/main.htm.

U.S. Department of Energy, Photovoltaic Program, http://www.eren.doe.gov/pv/text_frameset.html.

SOLAR POND

Department of Mechanical and Industrial Engineering

University of Texas at El Paso
El Paso, TX 79968.
E-mail: aswift@cs.utep.edu.

SPENT FUEL

Environmental Protection Agency, www.ntp.doe.gov, www.rw.doe.gov/pages/resource/facts/transfct.htm.

SUPERFUND

Environmental Protection Agency, http://www.epa.gov/epaoswer/hotline.

Recycled Superfund sites, http://www.epa.gov/superfund/programs/recycle/index.htm.

Superfund Information, http://www.epa.gov/superfund.

U.S. EPA Superfund Program Homepage, Website: http://www.epa.gov/superfund/index.htm.

TENNESSEE VALLEY AUTHORITY

Homepage, http://www.tva.gov.

THOREAU, HENRY

Website: http://www.walden.org.

TOXIC CHEMICALS

Environmental Defense Fund, http://www.scorecard.org.

U.S. Department of Health and Human Services, Agency for Toxic Substances and Disease Registry (ASTDR), http://www.atsdr.cdc.gov/

U.S. Environmental Protection Agency, Integrated Risk Information System (IRIS), http://www. siri.org/msds; http://www.ilpi.com/mads/index.html.

U.S. Occupation Health and Safety Administration, http://www.toxicsubstances/index.html.

TOXIC RELEASE INVENTORY

Environmental Defense Fund, http://www.scorecard.org.

Environmental Protection Agency, http://www.epa.gov.

Teach with Databases, Toxic Release Inventory, http://www.nsta.org/pubs/special/pb143x01.htm.

TOXIC WASTE

Environmental Defense Fund, http://www.scorecard.org.

Institute for Global Communications, http://www.igc.org/igc/issues/tw/.

TRADE RECORDS ANALYSIS OF FLORA AND FAUNA IN COMMERCE (TRAFFIC)

Homepage, http://www.traffic.org/about/.

URBAN FORESTS

American Forests, http://www.amfor.org.

TreeLink, http://www.treelink.org.

VERTEBRATES

U.S. Fish and Wildlife Service, Species List of Endangered and Threatened Wildlife, http://www.fws.gov/r9endspp/lsppinfo.html.

VICUNA

U.S. Fish and Wildlife Service, Species List of Endangered and Threatened Wildlife, http://endangered.fws.gov

VITRIFICATION

U.S. Department of Energy, http://www.em.doe.gov/fs/fs3m.html.

VOLCANOES

USGS, Volcanoes in the Learning Web, http://www.usgs.gov/education/learnweb/volcano/index.html.

Volcano Hazards, http://volcanoes.usgs.gov/.

WATER CONSERVATION AND POLLUTION

Early History of the Clean Water Act, http://epa.gov/history/topics.

Environmental Protection Agency, Office of Wetlands, Oceans, Watersheds for Nonpoint Source information, http://www.epa.gov/owow/wetlands/wetland2.html; http://www.epa.gov/swerosps/ej/.

U.S. Geological Survey, Water Resources of the United States, National Groundwater Association Homepage, http://www.h2o-ngwa.org.

Water Resources Information, http://water.usgs.gov/.

Water Use Data, http://water.usgs.gov/public/watuse/.

WETLANDS

National Wetlands Research Center, http://www.nwrc.usgs.gov/educ_out.html.

Ramsar Convention on Wetlands (International), http://www2.iucn.org/themes/ramsar/.

Ramsar List of Wetlands of International Importance, http://ramsar.org/key_sitelist.htm.

WHALES

Institute of Cetacean Research (ICR), http://www.whalesci.org.

U.S. Fish and Wildlife Service, Species List of Endangered and Threatened Wildlife, http://www.fws.gov/r9endspp/lsppinfo.html; http://www.highnorth.no/iceland/th-in-to.htm; http://greenpeace.org/.

WILDERNESS

U.S. Forest Service, *Roadless Area Review and Evaluation*, http://www.fs.fed.us.

Wilderness Society, http://www.wilderness.org/newsroom/factsheets.htm.

WILDLIFE REFUGES

Conservation International, http://www.conservation.org.

Nature Conservancy, http://www.tnc.org.

U.S. Fish and Wildlife Service, National Wildlife Refuge System, http://refuges.fws.gov.

World Conservation Union/International Union for the Conservation of Nature, http://www.iucn.org.

WIND ENERGY

American Wind Energy Association

122 C Street NW, 4th Floor
Washington, D.C. 20001
Telephone: (202) 383-2500.
E-mail: awea@mcimail.com.
Website: http://www.awea.org.

Center for Renewable Energy and Sustainable Technology (CREST)

Solar Energy Research and Education Foundation
777 North Capitol Street NE, Suite 805
Washington, D.C. 20002
Website: http://solstice.crest.org/.

WOLVES

U.S. Fish and Wildlife Service, http://www.fws.gov/.
U.S. Fish and Wildlife Service, Species List of Endangered and Threatened Wildlife, http://endangered.fws.gov/.

World Wildlife Fund

1250 24th Street, NW
Washington, D.C. 20037
Telephone: 1-800-225-5993
Website: http://www.worldwildlife.org/.

WORLD HEALTH ORGANIZATION

Homepage, http://www.who.int.

WORLD WILDLIFE FUND

1250 24th Street, NW
Washington, D.C. 20037
Telephone: 1-800-225-5993
Website: http://www.wwf.org/.

YUCCA MOUNTAIN PROJECT

Homepage, http://www.ymp.gov/.

ZEBRAS

U.S. Fish and Wildlife Service, Species List of Endangered and Threatened Wildlife, http://endangered.fws.gov/.

ZOOS

Bronx Zoo, http://www.bronxzoo.com/.
San Diego Zoo, http://www.sandiegozoo.org/.

APPENDIX D: ENVIRONMENTAL ORGANIZATIONS

Action for Animals

P.O. Box 17702
Austin, TX 78760
Telephone: (512) 416-1617
Fax: (512) 445-3454
Website: http://www.envirolink.org/

African Wildlife Foundation (AWF)

1400 Sixteenth Street, NW, Suite 120
Washington, D.C. 20036
Telephone: (202) 939-3333
Fax: (202) 939-3332
Website: http://www.awf.org/home.html

Agency for Toxic Substances and Diseases, Registry Division of Toxicology (ATSDR)

1600 Clifton Road
NE Mailstop E-29
Atlanta, GA 30333
Telephone: (888) 42-ATSDR or (888) 422-8737
E-mail: ATSDRIC@cdc.gov
Website: http://www.atsdr.cdc.gov/
contacts.html

Alaska Forum for Environmental Responsibility

P.O. Box 188
Valdez, AK 99686
Telephone: (907) 835-5460
Fax: (907) 835-5410
Website: http://www.accessone.com/~afersea

American Conifer Society (ACS)

P.O. Box 360
Keswick, VA 22947-0360
Telephone: (804) 984-3660
Fax: (804) 984-3660

E-mail: ACSconifer@aol.com
Website: http://www.pacificrim.net/~bydesign/
acs.html

American Forests

P.O. Box 2000
Washington, D.C. 20013
Telephone: (202) 955-4500
Website: http://www.americanforests.org

American Nuclear Society

555 North Kensington Avenue
La Grange Park, IL 60525
Telephone: (708) 352-6611
Fax: (708) 352-0499
E-mail: NUCLEUS@ans.org
Website: http://www.ans.org

American Oceans Campaign

201 Massachusetts Avenue NE, Suite C-3
Washington, D.C. 20002
Telephone: (202) 544-3526
Fax: (202) 544-5625
E-mail: aocdc@wizard.net
Website: http://www.americanoceans.org

American Rivers

1025 Vermont Avenue NW, Suite 720
Washington, D.C. 20005
Telephone: (202) 347-7500
Fax: (202) 347-9240
E-mail: amrivers@amrivers.org
Website: http://www.amrivers.org

American Society for Horticultural Science (ASHS)

600 Cameron Street
Alexandria, VA 22314-2562

Telephone: (703) 836-4606
Fax: (703) 836-2024
E-mail: webmaster@ashs.org
Website: http://www.ashs.org

American Society for the Prevention of Cruelty to Animals (ASPCA)

424 East Ninety-second Street
New York, NY 10128
Telephone: (212) 876-7700
Website: http://www.aspca.org

American Solar Energy Society

2400 Central Avenue, Suite G-1
Boulder, CO 80301
Telephone: (303) 443-3130
Fax: (303) 443-3212
E-mail: ases@ases.org
Website: http://www.ases.org
Publication: *Solar Today*

American Wind Energy Association

122 C Street NW, Fourth Floor
Washington, D.C. 20001
Telephone: (202) 383-2500
E-mail: awea@mcimail.com
Website: http://www.awea.org

Animal Legal Defense Fund (ALDF)

127 Fourth Street
Petaluma, CA 94952
Telephone: (707) 769-7771
Fax: (707) 769-0785
E-mail: info@aldf.org
Website: http://www.aldf.org

Animal Rights Network

P.O. Box 25881
Baltimore, MD 21224
Telephone: (410) 675-4566
Fax: (410) 675-0066
Website: http://www.envirolink.org/arrs/aa/
    index.html
Publication: *Animals' AGENDA*, a bimonthly
    magazine

Baron's Haven Freehold

104 South Main Street
Cadiz, OH 43907

Telephone: (740) 942-8405
Website: http://bhfi.1st.net

Biodiversity Support Program (BSP)

1250 North Twenty-fourth Street NW,
    Suite 600
Washington, D.C. 20037
Telephone: (202) 778-9681
Fax: (202) 861-8324
Website: http://www.BSPonline.org

Biosfera

Pres. Vargas 435, Suites 1104 and 1105
Rio de Janeiro, RJ 20077-900
Brazil

Birds of Prey Foundation

2290 South 104th Street
Broomfield, CO 80020
Telephone: (303) 460-0674
Fax: (303) 666-1050
E-mail: raptor@birds-of-prey.org
Website: http://www.birds-of-prey.org

Build the Earth

3818 Surfwood Road
Malibu, CA 90265
Telephone: (310) 454-0963

Center for Conversion and Research of Endangered Wildlife (CREW)

Cincinnati Zoo and Botanical Garden
3400 Vine Street
Cincinnati, OH 45220
E-mail: terri.roth@cincyzoo.org

Center for Marine Conservation

1725 DeSales Street SW, Suite 600
Washington, D.C. 20036
Telephone: (202) 429-5609
Fax: (202) 872-0619
E-mail: cmc@dccmc.org
Website: http://www.cmc-ocean.org

Centers for Disease Control (CDC)

1600 Clifton Rd.
Atlanta, GA 30333

Telephone: (800) 311-3435
Website: http://www.cdc.gov

Cheetah Conservation Fund (CCF)

P.O. Box 1380
Ojai, CA 93024
Telephone: (805) 640-0390
Fax: (815) 640-0230
E-mail: info@cheetah.org
Website: http://www.cheetah.org

Clean Air Council (CAC)

135 South Nineteenth Street, Suite 300
Philadelphia, PA 19103
Telephone: (888) 567-7796
Website: http://www.libertynet.org/
~cleanair/

Coalition for Economically Responsible Economies (CERES)

11 Arlington Street, Sixth Floor
Boston, MA 02116-3411
Telephone: (617) 247-0700
Fax: (617) 267-5400
Website: http://www.ceres.org

Conservation International

1015 Eighteenth Street NW Suite 1000
Washington, D.C. 20036
Telephone: (202) 429-5660
Website: http://www.conservation.org/
Publication: *Orion Nature Quarterly*

Convention on International Trade in Endangered Species of Wild Fauna and Flora (CITES)

CITES Secretariat
International Environment House,
15, chemin des Anémones, CH-1219
Châtelaine-Geneva, Switzerland
E-mail: cites@unep.ch
Website: http://www.cites.org/index.shtml

Council for Responsible Genetics

5 Upland Road, Suite 3
Cambridge, MA 02140
Website: http://www.gene-watch.org

Cousteau Society

870 Greenbriar Circle, Suite 402
Chesapeake, VA 23320
Telephone: (804) 523-9335
E-mail: cousteau@infi.net
Website: http://www.cousteausociety.org/
Publication: *Calypso Log*

Defenders of Wildlife

1101 Fourteenth Street NW, Room 1400
Washington, D.C. 20005
Telephone: (800) 441-4395
Website: http://www.Defenders.org
Publication: *Defenders*, a quarterly magazine

Dian Fossey Gorilla Fund International

800 Cherokee Avenue SE
Atlanta, GA 30315-1440
Telephone: (800) 851-0203
Fax: (404) 624-5999
E-mail: 2help@gorillafund.org
Website: http://www.gorillafund.org/
000_core_frmset.html

Earth Day Network

1616 P Street NW
Suite 200
Washington, D.C. 20036
E-mail: earthday@earthday.net
Website: http://www.earthday.net

Earth Island Institute (EII)

300 Broadway, Suite 28
San Francisco, CA 94133
Telephone: (415) 788-3666
Fax: (415) 788-7324
Website: http://www.earthisland.org/abouteii/
abouteii.html
Publication: *Earth Island Journal*, a quarterly
magazine

Earth, Pulp, and Paper

P.O. Box 64
Leggett, CA 95585
Telephone: (707) 925-6494
E-mail: tree@tree.org
Website: http://www.tree.org/epp.htm

EarthFirst! (EF!)

P.O. Box 5176
Missoula, MT 59806
Website: http://www.webdirectory.com/
    General_Environmental_Interest/
    Earth_First_/

Earthwatch Institute

In United States and Canada
3 Clocktower Place, Suite 100
Box 75
Maynard, MA 01754
Telephone: (800) 776-0188 or (617) 926-8200
Fax: (617) 926-8532
In Europe
57 Woodstock Road
Oxford OX2 6HJ, United Kingdom
E-mail: info@uk.earthwatch.org
Website: http://www.earthwatch.org

EcoCorps

1585 A Folsom Avenue
San Francisco, CA 94103
Telephone: (415) 522-1680
Fax: (415) 626-1510
E-mail: eathvoice@ecocorps.org
Website: http://www.owplaza.com/eco

Ecotourism Society

P.O. Box 755
North Bennington, VT 05257
Telephone: (802) 447-2121
Fax: (802) 447-2122
E-mail: ecomail@ecotourism.org
Website: http://www.ecotoursim.org

E. F. Schumacher Society

140 Jug End Road
Great Barrington, MA 01230
Telephone: (413) 528-1737
E-mail: efssociety@aol.com
Website: http://members.aol.com/efssociety/
    index.html

Electric Vehicle Association of the Americas

701 Pennsylvania Avenue NW, Fourth Floor
Washington, D.C. 20004
Telephone: (202) 508-5995
Fax: (202) 508-5924
Website: http://www.evaa.org

Environmental Defense Fund (EDF)

257 Park Avenue South
New York, NY 10010
Telephone: (800) 684-3322
Fax: (212) 505-2375
E-mail (for general questions and information):
    Contact@environmentaldefense.org
Website: http://www.edf.org
Publication: *Nature Journal*, a monthly
    magazine

Exotic Cat Refuge and Wildlife Orphanage

Route 3, Box 96A
Kirbyville, TX 75956
Telephone: (409) 423-4847

Federal Emergency and Management Agency (FEMA)

500 C Street SW
Washington, D.C. 20472
Website: http://www.fema.gov

Friends of the Earth (FOE)

1025 Vermont Avenue NW, Suite 300
Washington, D.C. 20005-6303
Telephone: (202) 783-7400
Fax: (202) 783-0444
E-mail: foe@foe.org
Website: http://www.foe.org

Green Seal

1001 Connecticut Avenue NW, Suite 827
Washington, D.C. 20036-5525
Telephone: (202) 872-6400
Fax: (202) 872-4324
Website: http://www.greenseal.org

Greenpeace USA

1436 U Street NW
Washington, D.C. 20009
Telephone: (202) 462-1177
Website: http://www.greenpeaceusa.org/
Publication: *Greenpeace Magazine*

Hawkwatch International

P.O. Box 660
Salt Lake City, UT 84110
Telephone: (801) 524-8511
E-mail: hawkwatch@charitiesusa.com
Website: http://www.vpp.com/HawkWatch

Humane Society of the United States (HSUS)

2100 L Street NW
Washington, D.C. 20037
Website: http://www.hsus.org
Publications: *All Animals*, a quarterly magazine

International Atomic Energy Commission

P.O. Box 100
Wagramer Strasse 5
A-1400, Vienna, Austria
E-mail: Official.Mail@iaea.org
Website: http://www.iaea.org

International Council for Local Environmental Initiatives (ICLEI)

World Secretariat
16th Floor, West Tower, City Hall
Toronto, M5H 2N2, Canada
Fax: (416) 392-1478
Email: iclei@iclei.org
Website: http://www.iclei.org

International Rhino Foundation (IRF)

14000 International Road
Cumberland Ohio 43732
E-mail: IrhinoF@aol.com
Website: http://www.rhinos-irf.org

International Whaling Commission (IWC)

The Red House
135 Station Road
Impington, Cambridge CB4 9NP,
    United Kingdom
E-mail: iwc@iwcoffice.org
Website: http://ourworld.compuserve.com/
    homepages/iwcoffice

International Wolf Center

1396 Highway 169
Ely, MN 55731-8129

Telephone: (218) 365-4695
Fax: (218) 365-3318
Website: http://www.wolf.org

Jane Goodall Institute (JGI)

P.O. Box 14890
Silver Spring, MD 20911-4890
Telephone: (301) 565-0086
Fax: (301) 565-3188
E-mail: JGIinformation@janegoodall.org

Keep America Beautiful

1010 Washington Boulevard
Stamford, CT 06901
Telephone: (203) 323-8987
Fax: (203) 325-9199
E-mail: info@kab.org

League of Conservation Voters

1707 L Street, NW, Suite 750
Washington, D.C. 20036
Telephone: (202) 785-8683
Fax: (202) 835-0491
E-mail: lcv@lcv.org
Website: http://www.lcv.org

Mountain Lion Foundation (MLF)

P.O. Box 1896
Sacramento, CA 95812
Telephone: (916) 442-2666
E-mail: MLF@moutainlion.org
Website: http://www.mountainlion.org

National Alliance of River, Sound, and Bay Keepers

P.O. Box 130
Garrison, NY 10524
Telephone: (800) 217-4837
E-mail: keepers@keeper.org
Website: http://www.keeper.org

National Anti-Vivisection Society (NAVS)

53 West Jackson Street, Suite 1552
Chicago, IL 60604
Telephone: (800) 888-NAVS
E-mail: navs@navs.org
Website: http://www.navs.org

National Arbor Day Foundation

100 Arbor Avenue
Nebraska City, NE 68410
Telephone: (402) 474-5655
Website: http://www.arborday.org
Publication: *Arbor Day*, a bimonthly magazine

National Audubon Society (NAS)

700 Broadway
New York, NY 10003
Telephone: (212) 979-3000
Website: http://www.audubon.org
Publication: *Audubon*, a bimonthly magazine

National Center for Environmental Health

Mail Stop F-29
4770 Buford Highway NE
Atlanta, GA 30341-3724
Telephone NCEH Health Line: (888)
232-6789
Website: http://www.cdc.gov/nceh/
ncehhome.htm

National Parks and Conservation Association (NPCA)

1015 Thirty-first Street NW
Washington, D.C. 20007
Telephone: (202) 944-8530; (800) NAT-PARK
E-mail: npca@npca.org
Website: http://www.npca.org
Publication: *National Parks*, a bimonthly
magazine

National Wildlife Federation (NWF)

8925 Leesburg Pike
Vienna, VA 22184-0001
Telephone: (800) 822-9919
Website: http://www.nwf.org
Publication: *National Wildlife*, a bimonthly
magazine

Natural Resources Defense Council (NRDC)

40 West Twentieth Street
New York, NY 10011
Website: http://www.nrdc.org
Publications: *Amiscus Journal*, a quarterly
magazine

Nature Conservancy (TNC)

1815 North Lynn Street
Arlington, VA 22209
Telephone: (703) 841-5300
Fax: (703) 841-1283
Website: http://www.tnc.org
Publication: *Nature Conservancy*, a magazine

Noise Pollution Clearinghouse

P.O. Box 1137
Montpelier, VT 05601-1137
Telephone: (888) 200-8332
Website: http://www.nonoise.org

North Sea Commission

Business and Development Office
Skottenborg 26, DK-8800 Viborg,
Denmark
Website: http:\\www.northsea.org

People for Animal Rights

P.O. Box 8707
Kansas City, MO 64114
Telephone: (816) 767-1199
E-mail: parinfo@envirolink.org
Website: http://www.parkc.org

People for the Ethical Treatment of Animals (PETA)

501 Front Street
Norfolk, VA 23510
Telephone: (757) 622-PETA
Fax: (757) 622-0457
Website: http://www.peta-online.org/

Orangutan Foundation International

822 South Wellesley Avenue
Los Angeles, CA 90049
Telephone: (800) ORANGUTAN
Fax: (310) 207-1556
E-mail: ofi@orangutan.org
Website: http://www.ns.net/orangutan

Ozone Action

1700 Connecticut Avenue NW, Third Floor
Washington, D.C. 20009
Telephone: (202) 265-6738

E-mail: cantando@essential.org
Website: www.ozone.org

Peregrine Fund

566 West Flying Hawk Lane
Boise, ID 83709
Telephone: (208) 362-3716
Fax: (208) 362-2376
E-mail: tpf@peregrinefund.org
Website: http://www.peregrinefund.org

Rachel Carson Council

8940 Jones Mill Road
Chevy Chase, MD 20815
Telephone: (301) 652-1877
E-mail: rccouncil@aol.com
Website: http://members.aol.com/rccouncil/
ourpage

Rainforest Action Network

221 Pine Street, Suite 500
San Francisco, CA 94104-2740
Telephone: (415) 398-4404
Fax: (415) 398-2732
E-mail: rainforest@ran.org
Website: http://www.ran.org

Range Watch

45661 Poso Park Drive
Posey, CA 93260
Telephone: (805) 536-8668
E-mail: rangewatch@aol.com
Website: http://www.rangewatch.org

Raptor Resource Project

2580 310th Street
Ridgeway, IA 52165
E-mail: rrp@salamander.com
Website: http://www.salamander.com~rpp

Reef Relief

201 William Street
Key West, FL 33041
Telephone: (305) 294-3100
Fax: (305) 923-9515
E-mail: reef@bellsouth.net
Website: http://www.reefrelief.org

ReefKeeper International

2809 Bird Avenue, Suite 162
Miami, FL 33133
Telephone: (305) 358-4600
Fax: (305) 358-3030
E-mail: reefkeeper@reefkeeper.org
Website: http://www.reefkeeper.org

Renewable Energy Policy Project-Center for Renewable Energy and Sustainable Technology (REPP-CREST)

National Headquarters
1612 K Street, NW, Suite 202
Washington, D.C. 20006
Website: http://www.solstice.crest.org

Resources for the Future (RFF)

1616 P Street NW
Washington, D.C. 20036
Telephone: (202) 328-5000
Fax: (202) 939-3460
E-mail: info@rff.org
Website: http://www.rff.org

Roger Tory Peterson Institute

311 Curtis Street
Jamestown, NY 14701
Telephone: (716) 665-2473
E-mail: webmaster@rtpi.org

Sierra Club

85 Second Street, Second Floor
San Francisco, CA 94105
Telephone: (415) 977-5630
Fax: (415) 977-5799
E-mail (general information):
information@sierraclub.org
Website: http://www.Sierraclub.org
Publication: *Sierra*, a bimonthly magazine

Smithsonian Institution Conservation & Research Center (CRC)

Website: http://www.si.edu/crc/brochure/
index.htm

Society of American Foresters

5400 Grosvenor Lane
Bethesda, MD 20814

Telephone: (301) 897-8720
Fax: (301) 897-3690
E-mail: safweb@safnet.org
Website: http://www.safnet.org

Surfrider Foundation USA

122 South El Camino Real, Suite 67
San Clemente, CA 92672
Telephone: (949) 492-8170
Fax: (949) 492-8142
Website: http://www.surfrider.org

Union of Concerned Scientists

National Headquarters
2 Brattle Square
Cambridge, MA 02238
Telephone: (617) 547-5552
E-mail: ucs@ucsusa.org
Website: http://www.ucsusa.org
Publications: *Nucleus*, a quarterly magazine;
   Earthwise, a quarterly newsletter

United Nations Environment Programme (Regional)

2 United Nations Plaza
NY, NY 10017
Telephone: (212) 963-8138
Website: http://www.unep.org

United Nations Food and Agriculture Organization (FAO)

Website: http://www.fao.org
Liaison office with North America
Suite 300, 2175 K Street NW, Washington D.C.
   20437-0001

United Nations Man and the Biosphere Programme (UNMAB)

U.S. MAB Secretariat, OES/ETC/MAB
Department of State
Washington, D.C. 20522-4401
Website: http://www.mabnet.org

U.S. Department of Agriculture (USDA)

14th Street and Independence Avenue., SW,
Washington, D.C. 20250
Website: http://www.usda.gov

U.S. Department of Energy (DOE)

Forrestal Building
1000 Independence Avenue, SW,
Washington, D.C. 20585
Website: http://www.doe.gov

U.S. Environmental Protection Agency (EPA)

401 M Street SW
Washington, D.C. 20460
Website: http://www.epa.gov

U.S. Fish and Wildlife Service (FWS)

1849 C Street NW
Washington, D.C. 20240
Telephone: (202) 208-5634
Website: http://www.fws.org

U.S. Geological Survey (USGS)

U.S. Dept. of Interior
1849 C Street, NW
Washington, D.C. 20240
Website: http://www.usgs.gov

U.S. National Park Service (NPS)

U.S. Dept. of Interior
1849 C Street, NW
Washington, D.C. 20240
Website: http://www.nps.gov

U.S. Nuclear Regulatory Commission (NRC)

One White Flint North
11555 Rockville Pike
Rockville, Maryland 20852
Website: http://www.nrc.gov

Wilderness Society

900 Seventeenth Street NW
Washington, D.C. 20006-2506
Telephone: (800) THE-WILD
Website: www.wilderness.org

Wildlands Project (TWP)

1955 West Grant Road, Suite 145
Tucson, AZ 85745
Telephone: (520) 884-0875
Fax: (520) 884-0962

E-mail: information@twp.org
Website: http://www.twp.org

World Conservation Monitoring Centre (WCMC)

219 Huntington Road
Cambridge CB3 ODL, United Kingdom
E-mail: info@wcmc.org.uk
Website: http://www.wcmc.org.uk

World Conservation Union (IUCN)

1630 Connecticut Avenue NW, Third Floor
Washington, D.C. 20009-1053
Telephone: (202) 387-4826
Fax: (202) 387-4823
E-mail: postmaster@iucnus.org
Website: http://www.iucn.org

World Health Organization (WHO)

Avenue Appia 20
1211 Geneva 27
Switzerland
Website: http://www.eho.int
E-mail: inf@who.int

World Parrot Trust United States

P.O. Box 50733
Saint Paul, MN 55150
Telephone: (651) 994-2581
Fax: (651) 994-2580
E-mail: usa@worldparrottrust.org

United Kingdom

Karen Allmann, Administrator,
Glanmor HouseHayle,
Cornwall TR27 4HY,
United Kingdom
E-mail: uk@worldparrottrust.org

Australia

Mike Owen
7 Monteray Street
Mooloolaba, Queensland 4557, Australia
E-mail: australia@worldparrottrust.org
Website: http://www.world parrottrust.org

World Resources Institute

1709 New York Avenue NW
Washington, D.C. 20006
Telephone: (202) 638-6300
E-mail: info@wri.org
Website: http://www.wri.org/wri/biodiv

World Society for the Protection of Animals (WSPA)

P.O. Box 190
Jamaica Plain, MA 02130
Website: http://www.wspa.org

United Kingdom Division
Website: http://www.wspa.org.uk/home.html

World Wildlife Fund, US (WWF)

1250 Twenty-fourth Street NW
P.O. Box 97180
Washington, D.C. 20077-7180
Telephone: (800) CALL-WWF
Website: http://www.worldwildlife.org

WorldWatch Institute

1776 Massachusetts Avenue NW
Washington, D.C. 20036
Telephone: (202) 452-1999
Website: http://www.worldwatch.org/
Publications: *WorldWatch, State of the World, Vital Signs* (annuals)

Zero Population Growth

1400 Sixteenth Street NW, Suite 320
Washington, D.C. 20036
Telephone: (202) 332-2200
Fax: (202) 332-2302
E-mail: zpg@igc.apc.org
Website: http://www.zpg.org

Zoe Foundation

983 River Road
Johns Island, SC 29455
Telephone: (803) 559-4790
E-mail: savage@awod.com
Website: http://www.2zoe.com

INDEX

f indicates figures and photos; t indicates tables

plant, **2:**103, **5:**48–49, **5:**49f
world, **2:**102t
Aquatic ecosystems, succession in,
  1:133–134, **1:**134f
Aquatic life, acid rain effects on, **4:**23f
Aqueduct(s), **2:**91
Aquifer(s), **2:**93, **2:**93f, **4:**43–45, **4:**43f,
  4:45f
  defined, **1:**99, **3:**49, **4:**43f, **4:**55
  depletion of, **4:**44–46, **4:**44t, **4:**45f
  in Great Plains, in America, **3:**41
  Ogallala, **2:**98, **2:**99f, **3:**41, **4:**45, **4:**45f
Arable, defined, **3:**31
Archaebacteria, **1:**19–20, **1:**19f, **1:**20f,
  1:20t
Archeological, defined, **2:**125
Archeological, defined, **2:**125
Arctic
  air pollution in, **4:**6–7, **4:**6f, **4:**7f
  carbon monoxide in, **4:**6–7, **4:**7f
Arizona's Petrified Forest, **2:**111
Arsenic, **4:**70–71
Art, in Stone Age, **3:**6
Artesian well, **2:**94, **2:**94f
Asbestos, **2:**71
  indoor air pollution by, **4:**12t, **4:**13
Asia, stone tools in, **3:**7
Assembly line, defined, **3:**72
Association of Consulting Foresters, **5:**58
Atmosphere, **1:**7–11, **1:**8f, **1:**9f, **1:**11f,
  4:18–19, **4:**19f
  climate, **1:**10–11, **1:**11f
  gases in, **1:**7–8
  greenhouse effect on, **1:**8, **1:**8f
  layers of, **1:**8–10, **1:**9f
Atom(s), defined, **2:**31
Atomic Energy Commission (AEC), **2:**26
Australia's Great Barrier Reef, **1:**116,
  1:117f
Automobile(s)
  fuel cell, **2:**52–53, **2:**52f, **2:**53f
  fuel cells for, **5:**31–33, **5:**32f, **5:**33f
  during Industrial Revolution, **3:**63–64,
  3:64f
  manufacturing of, eco-efficiency in,
  5:73, **5:**73f
AWEA. *See* American Wind Energy
  Association (AWEA)
Axe(s), hand, **3:**5–6, **3:**5f

Bacteria
  chemosynthetic, **1:**22–23
  coliform, **4:**40, **4:**41
  defined, **5:**51
  diseases caused by, **1:**23t
  round, **1:**23f
  spiral, **1:**23f
Bad breathing cities, **4:**1t
Bagasse, **2:**80, **5:**59
Bald eagle, **5:**68f, **5:**69t
Barter economy, defined, **3:**85
Bat(s), endangered, **4:**115t, **4:**116f
Battery(ies)
  INMETCO Recycling Facility's
  acceptance of, **5:**76t
  recycling of, by INMETCO recycling
  facility, **4:**65t
Bay of Fundy, **1:**115, **1:**115f
Beauty aids, waterway contamination
  due to, **4:**37

Becquerel, Edmond, **2:**42, **5:**25
Beef, as food source, **3:**95
Bench terracing, **5:**40, **5:**40f
Benthic zone, **1:**104
  defined, **1:**121
  of open ocean zone, **1:**119–120, **1:**121f
Benzene, **4:**72–23
Bering Sea, **2:**101, **2:**101f, **5:**35f
Bessemer process, defined, **3:**72
Beverage Container Act (Bottle Bill),
  4:66–67
Bike commuting, **5:**99, **5:**100f
Biocentrism, **5:**14
Biodegradable, defined, **4:**69
Biodiversity, **1:**65–66, **4:**109–111, **4:**111t
  defined, **1:**77, **4:**109, **5:**15
  protecting of, **5:**3, **5:**60–64, **5:**62t,
  5:63f, **5:**63t
  treaties, laws, and lists in, **4:**111–116,
  4:113f, **4:**113t, **4:**115f
  value of, **4:**110–111
Biodiversity Treaty, **4:**111, **5:**60
Biofuel(s), **2:**48–49, **5:**28
  environmental concerns of, **2:**49
Biogeochemical cycles, **4:**18
  defined, **4:**34
Biological control, **5:**41
  defined, **5:**51
Biological diversity, changes in,
  deforestation and, **4:**93–94
Biomass, **2:**47–49, **5:**27–28
  defined, **1:**54
  described, **2:**47–48, **5:**27–28
Biome(s)(s), **1:**18, **1:**18f
  average annual rainfall of, **1:**58f, **1:**70f,
  1:81f, **1:**92f, **1:**97f
  defined, **2:**72, **4:**101
  freshwater. *See* Freshwater biomes
  land, **1:**56–78. *See also* Forest(s)
  marine, **1:**109–122. *See also* Marine
  biomes
  water, **1:**101–122. *See also* Water biomes
Biosphere, **1:**2–3, **1:**2f, **1:**17–39
  components of, **1:**2
  defined, **1:**2
  described, **1:**2, **1:**2f
  size of, **1:**2
Biosphere Reserves, **5:**65
Biotechnology, described, **5:**44–46, **5:**45t
Biphenyl(s), polychlorinated, **4:**73
Bird(s), **1:**27
  dodo, **4:**103, **4:**104f
Birthrate, measuring of, **3:**88
Bison
  in Great Plains, **3:**42f
  in Yellowstone National Park, **2:**115f
Black Death, defined, **3:**100
"The Black Land," **3:**18
Blue Angel, **5:**78
Bog(s), **1:**14, **1:**106, **1:**107f
Boiling water reactor (BWR), **2:**25
Bone tools, in New Stone Age, **3:**7f, **3:**8,
  3:8f
Bottle Bill, **4:**64–66
Bottle gas, **2:**7
Bowhead, **1:**119f
Breaker boys, **3:**58
Breeding
  livestock, **3:**37
  selective, defined, **3:**49
Brick(s), in Mesopotamia, **3:**16
British Thermal Units (BTUs), **2:**12
Bronze Age, **3:**15
Brower, David, **5:**13
Brownfield(s), **4:**79, **4:**79f, **4:**80f
  cleaning up of, **5:**89–90, **5:**89–91f

defined, **5:**102
  Swedish housing on, **5:**91, **5:**91f
Brundtland Report, **5:**9–10
BTUs. *See* British Thermal Units
  (BTUs)
Bullard, Linda McKeever, **5:**109
Bullard, Robert, **5:**109, **5:**110f
Bullitt Foundation, **5:**113, **5:**113f
Bureau of Land Management (BLM),
  2:109, **2:**119–121, **2:**119f–121f
Bus(es), **5:**96, **5:**97f
  cleaner fuels for, **5:**96–97, **5:**97f, **5:**98f
  electric, **5:**97f
  propane, **5:**97f
Business stewardship
  eco-efficiency in, **3:**85, **3:**85f, **5:**71–72.
  See also Eco-efficiency
  eco-labeling in, **5:**78–79
  sustainable, **5:**71–84. *See also* Sustainable
  business stewardship
BWR. *See* Boiling water reactor (BWR)

CAA. *See* Clean Air Act (CAA)
Clean Air Act, **1:**9
Cactus(i), **1:**91–92, **1:**93f, **1:**94
  endangered, **4:**113f, **4:**113t
Cadmium, **4:**73
  defined, **2:**31
Calcium cycle, **1:**54
California condor, **4:**119f, **4:**120t
Canada, stone tools in, **3:**7
Canada Deuterium Uranium (Candu)
  reactor, **2:**25
Candu reactor, **2:**25
Canopy, defined, **1:**77
Cape Hatteras National Seashore, **2:**111
Capital, defined, **3:**85
Capital resources
  defined, **5:**83
  production of, **3:**76–77
Capitalism, defined, **3:**81, **3:**85
Captive propagation, **4:**117–120,
  4:118f–120f, **4:**118t
  defined, **5:**69
Captive propagation programs, **5:**65
Cara Biological Reserve, **5:**66, **5:**67
Carbohydrate, defined, **1:**54
Carbon, in environment, **1:**50–51
Carbon cycle, **1:**50–52, **1:**51f, **4:**27f
Carbon dioxide, **4:**26–27, **4:**27f
  atmospheric concentrations of, **5:**11f
  defined, **5:**33
  global warming due to, **5:**58
  sources of, **4:**29f, **5:**17, **5:**17f
Carbon dioxide molecule, **4:**7f
Carbon monoxide, in Arctic, **4:**6–7, **4:**7f
Carbon monoxide molecule, **4:**7f
Carcinogen(s), defined, **4:**87
Caribbean Plate, **1:**4, **1:**5f
"Caring for the Land and Serving
  People," **2:**115
Carnivore(s), in ecosystem, **1:**42–43, **1:**43f
Carrying capacity
  defined, **3:**100, **4:**101, **5:**15
  of population, **3:**90–91
Carson, Rachel Louise, **5:**114–115
Cartwright, Edmund, in Industrial
  Revolution in America, **3:**53–54
Catalyst
  defined, **4:**16, **5:**33
  described, **5:**30
Catalytic converters, in air pollution
  control, **4:**10
Cat's Claw, **2:**83, **2:**83f
Cell(s)
  defined, **2:**105

Feldspar, **2:**65
Fertile Crescent, **3:**15, **3:**15f
  legacy of, **3:**17
Fertilizer(s)
  history of, **3:**36
  natural, **3:**35
  pollution due to, **4:**39
Fertilizer runoff, agricultural pollution
    due to, **4:**95
FIFRA. *See* Federal Insecticide,
    Fungicide, and Rodenticide Act
    (FIFRA)
Finch species, on Galapagos Islands,
    1:136f
Finland, stone tools in, **3:**7
Fire(s), in Stone Age, **3:**3–4
Firewood, **5:**28
Fish, **1:**29, **1:**29f
  eco-labeling of, **5:**50, **5:**50f
Fish and Wildlife Service (FWS), **2:**109,
    4:112–113, **4:**117, **5:**61, **5:**64, **5:**65
Fish and Wildlife Service (FWS)
    Endangered Species, **5:**63
Fish aquaculture, **2:**100–102,
    2:101f–102f
Fish stocks, rebuilding of, **5:**49–50
Fishing
  commercial, **2:**100t, **5:**46, **5:**46t
  oceanic
    environmental concerns of, **2:**99–100
    as food source, **3:**95
  sustainable, **5:**46–51, **5:**46t, **5:**46t, **5:**49f,
      5:50f. *See also* Sustainable fishing
Fission, described, **2:**22f
Flint, **3:**3
Flint tools, **3:**3–4
Flood(s), ecosystem effects of, **1:**124
Flood irrigation, **2:**95–96
Floodplain, defined, **1:**121, **3:**32
Florida Everglades, **5:**114
Florida Everglades National Park,
    defined, **5:**117
Florida panther, **5:**62t, **5:**63f
Florida's Pelican Island, **4:**117
Fluidized bed combustion, reducing of,
    4:23–24
Fluoride, defined, **2:**105
Foliage, defined, **1:**140
Food
  electricity and, during Industrial
      Revolution, **3:**66, **3:**67f
  from forests, **2:**80–81
  production and distribution of, **3:**94–98
  sources of, **3:**94–96
    crops, **3:**94–95
    livestock, **3:**95
    oceanic fishing and aquaculture,
        3:95–96
Food and Agriculture Organization
    (FAO), of U.N., **5:**45–46, **5:**47, **5:**54
Food and Drug Administration (FDA),
    4:41
Food chain, **1:**44, **1:**45f
Food distribution, electricity and,
    during Industrial Revolution, **3:**66,
    3:67f
Food sources
  in Mesopotamia, **3:**15
  from ocean, **2:**98, **2:**99f

Food webs, **1:**46–46, **1:**47f
Ford, **5:**33
Ford, Henry, **3:**83, **5:**73f
Forest(s), **1:**56–78, **2:**74–88
  acid rain effects on, **4:**22
  benefits of, **5:**54–56
  contributions to, **2:**75–86
  described, **2:**74
  disappearing, **3:**47f, **5:**54–56, **5:**54f,
      5:55f
  Eastern and Western hemispheres, **1:**57f
  environmental concerns of, **2:**86–87
  fuelwood from, **2:**75–76, **2:**75f, **2:**76t
  land surface covered by, **5:**53
  loss of, **4:**89–102, **5:**54. *See also*
      Deforestation
  prospect of, **5:**53
  management of, **2:**81
  medicinal products from, **2:**81–86,
      2:82f–85f
  non-timber products from, **2:**80–81
  old growth, defined, **2:**87
  old-growth, **1:**75
  paper products from, **2:**78–79, **2:**79f,
      2:80f
  rainforests, **2:**83–86, **2:**83f–85f. *See
      also* Rainforest(s)
  temperate, **1:**66–67. *See also* Temperate
      rain forests
  sustainability of, management of,
      5:56–59, **5:**57f, **5:**59f
  sustainable, **5:**53–59. *See also* Sustainable
      forests
  taiga, **1:**56–61, **1:**56f–61f. *See also* Taiga
      forests
  temperate, **1:**61–68, **1:**62f–65f, **1:**67f.
      See also Temperate forests
  in Eastern and Western hemispheres,
      2:74f
  environmental concerns of, **1:**65–66
  tropical rain, **1:**68–75, **1:**69f–74f
  urban. *See* Urban forests
    in U.S., **2:**75f
  prevalence of, **5:**54
  wood products from, **2:**77–78, **2:**78t
Forest fires, ecosystem effects of, **1:**124,
    1:124f
Forest Service in 1905, **2:**115
Forest Stewardship Council (FSC), **5:**58
Fossil fuel(s), **2:**1–18. *See also specific types*
  coal resources, **2:**11–17, **2:**13f–15f,
      2:17t
  described, **2:**1
  gasoline, **2:**6–9
  natural gas, **2:**9–11, **2:**10t, **2:**11f
  petroleum, **2:**3–6, **2:**4f–6f
  resources from, **2:**1–9, **2:**2t, **2:**3f–6f,
      2:8t
Fossil fuel emissions, environmental
    effects of, **5:**17–18
Fractional distillation, **2:**6
  defined, **2:**18
Franklin, B., as paper merchant, **2:**79
Free enterprise system, **3:**80–81, **3:**81f
Fresh Kills landfill, **4:**62
Freshwater
  sources of, **2:**91–94, **2:**92, **2:**93f, **2:**94f,
      4:37f
  uses of, **2:**89–98, **2:**90f, **2:**91f, **2:**92t,
      2:93f, **2:**95t, **2:**97f, **2:**97t, **2:**98f
Freshwater biomes, **1:1:**101f, **1:**101,
    1:102–109, **1:**103t, **1:**104f–107f
  bogs, **1:**106, **1:**107f
  freshwater swamps, **1:**105–106
  lakes, **1:**104, **1:**104f
  marshes, **1:**107–108

ponds, **1:**103
rivers, **1:**102–103, **1:**103t
wetlands, **1:**105–108, **1:**107f
Freshwater bodies, **1:**13–14, **1:**14f
Freshwater marshes, **1:**14
Freshwater pollution, **4:**36–40, **4:**37f,
    4:37t, **4:**38f, **4:**39t
  agricultural pollution—related, **4:**37–40
  potable water—related, **4:**36–37
  thermal pollution—related, **4:**40
Freshwater swamps, **1:**105–106
Friends of McKinley, Inc., **5:**108f
FSC. *See* Forest Stewardship Council
    (FSC)
FTC. *See* Federal Trade Commission
    (FTC)
Fuel(s)
  cleaner, for buses, **5:**96–97, **5:**97f,
      5:98f
  fossil. *See* Fossil fuels; *specific types*
  nuclear, **2:**23
  soybean, **5:**27f
  spent, **2:**28
Fuel cell(s)
  for automobiles, **5:**31–33, **5:**32f, **5:**33f
  electrolytes in, **5:**30
  history of, **5:**33
  sources of, **5:**31f
Fuel cell automobiles, **2:**52–53, **2:**52f,
    2:53f
Fuel Cell Hybrid Vehicle (FCHV),
    Toyota's, **2:**52f
Fuel rods
  defined, **4:**87
  in nuclear reactors, **2:**23–24
Fuelwood, **2:**48
  countries with, **2:**76, **2:**76t
  environmental concerns of, **2:**48, **2:**76
  from forests, **2:**75–76, **2:**75f, **2:**76t
  historical background of, **2:**75–76
  during Industrial Revolution, **3:**56–57
Fungus(i), **1:**20t, **1:**25–26, **1:**25f
  biodiversity of, **5:**4t
Furrow irrigation, **2:**96
Fusion
  described, **2:**19, **2:**30, **2:**30f
  nuclear, **2:**29–30, **2:**30f
FWS. *See* Fish and Wildlife Service
    (FWS)

Gaia, described, **1:**2–3
Gaia hypothesis, **1:**2–3
Garbage, **4:**57
Garbage disposal, in Curitiba, Brazil,
    4:66
The Garden Project, **5:**105f
Gas(es)
  atmospheric, **1:**7–8
  coal, during Industrial Revolution, **3:**59
  greenhouse, **4:**26–27, **4:**27f
  natural
    during Industrial Revolution, **3:**59,
        3:60f
    liquefied. *See* Liquefied natural gas
        (LNG)
  soil, **2:**59–60
Gasoline, **2:**6–9
  described, **2:**6–7
Gateway National Recreation Area,
    2:112
GDP. *See* Gross domestic product (GDP)
Gems, valuable, **2:**69
General Motors, **5:**31, **5:**33, **5:**71
Genetic(s), defined, **3:**49, **4:**121
Geologists, defined, **1:**15
Geothermal, defined, **2:**125

Hydrogen fuel cells, **2:**51–53, **2:**52f, **2:**53f, **5:**30–31, **5:**31f
  concerns related to, **2:**53
  described, **2:**51–52
  source of, **2:**52f
Hydrological cycle, **1:**49f, **1:**50
Hydropower. *See* Hydroelectric power
Hydrosphere, **1:**11–15, **1:**12f–14f
  defined, **1:**11
  freshwater bodies, **1:**13–14
  ocean currents, **1:**12–13, **1:**13f
  oceans, **1:**11–12, **1:**12f
  wetlands, **1:**14–15
Hypothermia, defined, **4:**55
Hypothesized, defined, **1:**15

Ice age, defined, **1:**99
IFQs. *See* Individual fishing quotients (IFQs)
Ignitable, defined, **4:**71t
Immigration, defined, **3:**100
In situ, defined, **2:**72
Incineration, **4:**61–62
  in hazardous waste disposal, **4:**77–78, **4:**78f
Incinerator(s), **4:**62–63, **4:**62f
Income inequality, by country, **3:**99t, **5:**5t
India
  ancient, **3:**14f, **3:**30–31, **3:**30f
  climate of, **3:**30, **3:**30f
  early agriculture in, **3:**31
  early legacy of, **3:**31
  geography of, **3:**30
  monsoons in, **3:**30, **3:**30f
  natural resources of, human impact on, **3:**31
  stone tools in, **3:**6
  tigers in, **4:**103
  wind plants of, **2:**36
Indian-Australian Plate, **1:**4, **1:**5f
Indigenous, defined, **2:**87, **3:**11, **5:**15
Indigenous peoples, **3:**3, **3:**4f
  human rights for, **5:**6, **5:**6f
  of rainforest, **2:**86
Individual fishing quotients (IFQs), **5:**49
Individual transferable quotas (ITQs), **5:**50
Indoor pollutants. *See* Air pollutants, indoor
Industrial Revolution, **3:**50–72
  accidents in mines and factories during, **3:**68–69
  achievements during, **3:**69
  in America, **3:**53–56, **3:**54f, **3:**55f
    Edmund Cartwright in, **3:**53–54
    factory system, **3:**54–56, **3:**55f
    Samuel Slater in, **3:**53, **3:**54f
  benefits of, **3:**69
  defined, **2:**18
  electricity during, **3:**65–66, **3:**67f
    communications, **3:**65–66
    food and food distribution, **3:**66, **3:**67f
    light and power, **3:**65
    newspapers, **3:**66
  energy source during
    changes in, **3:**56–61, **3:**57f, **3:**58f, **3:**60f, **3:**61f
    charcoal, **3:**56–57
    coal, **3:**56–57, **3:**58f
    coal gas, **3:**59

fuel wood, **3:**56–57
  iron, **3:**57, **3:**59
  natural gas, **3:**59, **3:**60f
  steel, **3:**57, **3:**59
  waterpower, **3:**56, **3:**57f
  wind power, **3:**56
in England, **3:**51–53, **3:**53t
  birth of, **3:**52
  factory system, **3:**51–52
  petroleum resources, **3:**59–60, **3:**61f
fossil fuel demands during, **2:**1
health and medicine in, **3:**67–69
  health conditions during, **3:**68
  social, cultural, and environmental impact of, **3:**70
  timeline of (1701–1909), **3:**70–71
  transportation during, **3:**61–65, **3:**62f, **3:**64f
    automobiles, **3:**63–64, **3:**64f
    new roads, **3:**64–65
    railroads, **3:**63
    steam engines, **3:**61–62
    steam locomotives, **3:**62, **3:**62f
    steam power, **3:**61–62
    steamships, **3:**62
Industrial smog, **4:**7–8
Inequality, income, by country, **5:**5t
Influenza virus, epidemics of, population decline due to, **3:**92–93, **3:**92f
Infrastructure, **5:**102
  defined, **2:**53
INMETCO Recycling Facility, batteries accepted by, **4:**65t, **5:**76t
Inorganic, defined, **2:**72
Inorganic material, defined, **4:**69
Insectivore(s), defined, **1:**38
Integrated pest management (IPM), **5:**40–41, **5:**41f, **5:**42f
Interface Flooring Systems, **5:**76
International Agency for Research on Cancer, **2:**9
International Engine of the Year Awards, **5:**32
International reserves, **4:**117, **5:**65–67, **5:**67f, **5:**68f, **5:**69t
Intertidal zones, **1:**113–114
  life in, **1:**113–114
Invertebrate(s), **1:**27, **1:**27f
Ion(s), defined, **2:**73
IPM. *See* Integrated pest management (IPM)
Ireland, stone tools in, **3:**7
Iron
  during Industrial Revolution, **3:**57, **3:**59
  in Mesopotamia, **3:**16
Irrigated area
  in countries (1994), **2:**97t
  in United States, top (1997), **2:**95t
Irrigation
  in ancient Egypt, **3:**18–19
  drip, **2:**96–97, **2:**98f, **5:**42–43, **5:**43f
  environmental concerns of, **2:**97
  flood, **2:**95–96
  furrow, **2:**96
  methods of, **2:**95–97, **2:**97f
  pivot, **2:**97f, **5:**43f
  with saltwater, **5:**44
  water for, **2:**95–98, **2:**95t, **2:**97f, **2:**97t, **2:**98f
Isotope(s), defined, **2:**31
Itaipú Hydroelectric Power Plant, **2:**38f, **5:**29f
ITQs. *See* Individual transferable quotas (ITQs)
IUCN. *See* World Conservation Union (IUCN)

Jackson, Simon, **5:**109
John Deere plow, **3:**42, **3:**43f
J-shaped curve, population-related, **3:**91–92, **3:**91f
Julian, Percy, **5:**45f

Kayapo, **2:**86
Kenaf, **5:**59, **5:**59f
Kerosene, **2:**7
Keystone species, **1:**30
Kilowatt (kW), defined, **2:**53
Kimberly Clark papermill, **5:**59
Kinetic energy, **2:**34
  defined, **2:**53, **5:**33
Kingdom(s), classification of, **1:**19, **1:**20t
Kudzu, **4:**107
Kyoto Protocol, **4:**34, **5:**10, **5:**12f

Lake(s), **1:**13–14, **1:**104, **1:**104f
  environmental concerns of, **1:**105
Lake Baikal, **1:**106, **1:**106f
Land biomes, **1:**56–78. *See also* Forest(s)
Land degradation, causes of, **5:**3f
Land enclosure, **3:**35
Land Ordinance of 1785, **2:**119
Land resources, **2:**56–73
  forests, **2:**74–88
  minerals, **2:**64–72, **2:**67t, **2:**70f, **2:**71f
  phosphorus, **2:**69–70, **2:**70f
  soil, **2:**56–64, **2:**57f–59f, **2:**61f, **2:**63f, **2:**64f
Landfill(s), **4:**59–62, **4:**61f, **4:**62f
  described, **4:**59–60
  design of, **4:**60–62, **4:**61f
  Fresh Kills, **4:**62
  hazardous wastes in, **4:**72t
  PCB, **5:**111
  prevalence of, **4:**62
  sanitary, **4:**59–62, **4:**61f, **4:**62f
  secured, in hazardous waste disposal, **4:**77
Landsat satellite, photograph of Earth's surface by, **1:**1, **1:**1f
Landslide, defined, **1:**140
Larderello geothermal field, **2:**45, **5:**27
Latitude, **1:**10, **1:**10f
Lava, **1:**125
  defined, **1:**140
Leach, defined, **4:**34
Leachate, defined, **4:**87
Leaching, defined, **4:**34
Lead, **2:**66–68
Leaf(ves)
  maple, **1:**63f
  oak, **1:**63f
Lee, Charles, **5:**111
Legume(s), defined, **5:**52
Less-developed countries, population growth in, **3:**89
Levee, defined, **3:**32
Lichen(s), **1:**32.**1:**33f, **2:**57f
  defined, **1:**77, **4:**16
  types of, **4:**9f
Life cycle, of product, eco-efficiency and, **5:**72–77, **5:**73f, **5:**75f, **5:**76f
Light, electricity and, during Industrial Revolution, **3:**65
Light rail transit (LRT), **5:**97–98
Light water reactor (LWR), **2:**24–25
Lignite, **2:**12
Limestone, in ancient Egypt, **3:**20
Liming, **4:**24f
Lion(s), in Serengeti National Park, **5:**67f
Liquefied natural gas (LNG), **5:**97
  defined, **5:**102
Liquefied petroleum gas (LPG), **2:**7
Liquid, volatile, defined, **4:**101

National Soil Survey Center (NSSC),
2:63
National Trails System, 2:121, 2:124–125
National Trails System Act of 1968,
2:111–112
National Wild and Scenic Rivers
System, 2:121, 2:122t
National Wildlife Refuge System,
2:107–108
National Woodland Owners Association,
5:58
Native American(s), in Great Plains, 3:41
Native American wind energy plant,
2:33f
Native species, 1:19
Natural disasters
defined, 3:100
earthquakes, 1:124–125, 1:125f, 1:127t
ecosystem effects of, 1:123–130,
1:123f–126f, 1:127t, 1:128f, 1:129f
floods, 1:124
forest fires, 1:124, 1:124f
hurricanes, 1:127–128, 1:128f
monsoons, 1:128–130, 1:129f
typhoons, 1:128
volcanoes, 1:125–126, 1:125f, 1:126f
Natural fertilizers, 3:35
Natural gas, 2:9–11, 2:10t, 2:11f
contents of, 2:11
deposits of, locating of, 2:10–11
history of, 2:9
during Industrial Revolution, 3:59,
3:60f
liquefied. See Liquefied natural gas
(LNG)
liquid, 2:10
reserves of, by country, 2:10t
uses of, 2:10, 2:11
Natural resource(s)
defined, 3:86
production of, 3:76, 3:75t, 3:76f
Natural Resource Conservation Service,
2:60
Nature Conservancy, 5:87
NECAR 4, 5:32, 5:33
Needle-leaf trees, 1:59f
Nelson, Gaylord, 5:113
Neolithic period, 3:9–11, 3:9t, 3:10f,
3:11t
cultivation of wild plants in, 3:9–10,
3:9t, 3:10f
domestication of animals, 3:10–11, 3:11t
Neolithic revolution, 3:9
defined, 3:11
Neritic zone, 1:114–116, 1:114f, 1:115f
New Stone Age, technology in, 3:7–8,
3:7f, 3:8f
Newspaper(s), electricity and, during
Industrial Revolution, 3:66
NGOs. See Nongovernmental
organizations (NGOs)
Niche(s), 1:37–38, 1:37f
Nile River, 3:14f, 3:18, 3:20
Nitrification, 1:53
defined, 1:55
Nitrogen, 1:8
in DNA, 1:52
Nitrogen cycle, 1:52–53, 1:52f
in soil, 2:62, 2:64f

Nitrogen oxides, 4:4, 4:4t
emissions from, sources of, 4:4t
NOAA. See National Oceanic and
Atmospheric Administration (NOAA)
Nocturnal, defined, 1:99
Noise Control Act, 4:15
Noise levels, reducing of, 4:15–16
Noise pollution, 4:15–16, 4:16f
Nongovernmental organizations
(NGOs), 5:14
in wildlife protection, 5:68
Nonmetallic minerals, 2:69–72, 2:70f,
2:71f
Nonrenewable resources, defined, 4:69
Nonseed plants, 1:26
North American Plate, 1:4, 1:5f
Northwest Ordinance of 1787, 2:119
Northwest Rural Public Power District,
5:24
No-till farming, 5:37–38, 5:38f
NRC. See Nuclear Regulatory
Commission (NRC)
NSSC. See National Soil Survey Center
(NSSC)
Nuclear energy, 2:19–32
basics of, 2:19, 2:22
conversion to other forms of energy,
2:22
future of, 2:30–31
usage of, 2:26–29, 2:28t, 2:29f
history of, 2:21
world usage of, 2:19, 2:20t, 4:86t
Nuclear fuel, 2:23
Nuclear fusion, 2:29–30, 2:30f
Nuclear plants, in U.S., spent fuel stored
at, 4:82t
Nuclear power
advantages of, 2:26–27
environmental concerns related to,
2:27, 2:29
Nuclear power plants, 2:20–26, 2:23f,
2:25f. See also Nuclear reactor(s)
accidents from, 4:86–87
components of, 2:22–26
cooling towers of, 2:23f
in U.S., 2:19, 2:22f
world electrical generating capacity of,
2:31f
Nuclear power station sites, 2:28t
Nuclear reactor(s), 2:22–26
boiling water reactor, 2:25
Candu reactor, 2:25
components of, 2:21–24, 2:23f
control rods in, 2:24
coolant in, 2:24
fast breeder reactor, 2:25–26
fuel in, 2:23
fuel rods in, 2:23–24
function of, 2:22
light water reactor, 2:24–25
moderator in, 2:24
pressurized water reactor, 2:24
types of, 2:24–26, 2:25f
uranium in, 2:23
Nuclear Regulatory Commission
(NRC), 4:80–81
defined, 4:87
U.S., 2:26
Nuclear Waste Policy Act, 2:28
Nucleus(i), defined, 2:31
Nutrition, human, 3:94

Oak leaves, 1:63f
Obsidian, 3:3
Occupational Safety and Health
Administration (OSHA), 4:12

Ocean(s), 1:11–12, 1:12f
food sources from, 2:99
oil pollution in
laws to protect oceans, 4:54–55
sources of, 4:54, 4:54t
uses of, 2:98–105
Ocean currents, climate effects of,
1:12–13, 1:13f
Ocean Dumping Act, 4:54–55
Ocean fishing, environmental concerns
of, 2:99–100, 2:100t
Ocean minerals, 2:103–104
Ocean salts, composition of, 1:12f
Ocean thermal energy conversion
(OTEC), 2:49–51, 2:50f, 2:51f
described, 2:49–50
environmental concerns of, 2:50–51
of OTEC, 2:50–51
Ocean water, composition of, 1:109,
1:109t
Ocean water pollution, 4:48–52,
4:49f–51f
coral reefs effects of, 4:49, 4:49f, 4:50f
mangroves effects of, 4:50f
salt marshes effects of, 4:52
sources of, 4:48–49
Ocean zones, 1:114–116, 1:114f, 1:115f
Oceanic fishing, as food source, 3:95
Oceanographers, defined, 1:15
Octane, defined, 2:18
Octane number, 2:7
Ogallala Aquifer, 2:98, 2:99f, 3:41, 3:45,
4:45f
Oil(s)
extracting of, 2:5–6, 2:5f, 2:6f
petroleum, during Industrial
Revolution, 3:59–60, 3:61f
products made from, 2:7–9, 2:8t
Oil pollution, 4:52–54, 4:52f, 4:53t,
4:54t
in oceans
laws to protect oceans, 4:54–55
sources of, 4:54, 4:54t
oil spills, major, 4:53t
Oil reserves, by country, 2:2t
Oil spills, major, time capsule, 4:53t
Old growth, defined, 1:77
Old growth forests, 1:75
defined, 2:87
Old Stone Age, 3:2
Olmsted, Frederick Law, 5:94
Olympic Rain Forest, 1:66
Omnivore(s), in ecosystem, 1:43
Opaque, defined, 3:11
Open ocean zone, 1:117–120,
1:119f–121f
benthic zone, 1:119–120, 1:121f
deep zone, 1:119–120, 1:121f
surface zone, 1:118, 1:120f
Ore
defined, 2:73
mining of, in tons (1991), 3:78, 3:79t
Organic, defined, 2:73
Organic farming
defined, 5:52
sustainable agriculture vs., 5:36
Organic material, defined, 4:69
Organic solvent, defined, 4:87
Organism(s)
defined, 1:18
living. See Living organisms
Ornament(s), in New Stone Age, 3:8,
3:8f
Oryx, 4:120f
OSHA. See Occupational Safety and
Health Administration (OSHA)

Pulpwood, substitute for, **2:**80
Pumice, defined, **1:**140
PWR. *See* Pressurized water reactor
 (PWR)

Quartz, **2:**65
Quartzite, **3:**3, **3:**4f
Quebec Ministry of Environment, **4:**24
Quinine, **2:**83, **2:**84f
 source of, **5:**55f
 uses of, **5:**55f

Racism, environmental, **5:**4, **5:**5
 defined, **5:**15
Radiation
 adaptive, **1:**135
 health effects of, **4:**85–86
Radiation sickness, **4:**85–87, **4:**86t
Radioactive wastes
 disposal of, **4:**82–85, **4:**83f–85f
 high-level, **4:**81
 low-level, **4:**80–81
 nuclear, **4:**80–87
 radiation sickness due to, **4:**85–87, **4:**86t
 sources of, **4:**81–82
 transuranic wastes, **4:**81
 uranium mill tailings, **4:**82
Radon, indoor air pollution by, **4:**12t,
 4:13–14, **4:**14f
Rail transit, **5:**97–99
Railroad(s), during Industrial
 Revolution, **3:**63
Rain, acid. *See* Acid rain
Rainbow Bridge National Monument,
 2:121f
Rainforest(s), **1:**65f, **1:**66
 in Eastern and Western hemispheres,
 2:82f
 indigenous people of, **2:**86
 trees in, **2:**83–86, **2:**83f–85f
Rainforest Action Network (RAN),
 1:76, **5:**110, **5:**111f
Rainforest Alliance, **5:**7
Rainforests, temperate, **1:**66–67
RAN. *See* Rainforest Action Network
 (RAN)
RAN's Protect-an-Acre program, **1:**76
RCRA. *See* Resource Conservation and
 Recovery Act (RCRA)
Reactive, defined, **4:**71t
Reactor core, defined, **2:**31
Rechargeable Batteries Recycling
 Corporation, **5:**75–76, **5:**75f, **5:**76t
Recycled waste, products made from,
 4:64–65, **4:**65t
Recycling, **4:**63–67, **4:**64f, **4:**65t
 in hazardous waste disposal, **4:**78–79,
 4:79f, **4:**80f
"The Red Land," **3:**18
Red List of Endangered Species,
 4:113–116, **4:**113f–115f, **4:**113t,
 4:115t
Red List of Threatened and Endangered
 Species, **5:**62, **5:**63–64, **5:**63f, **5:**63t
Red tide, **4:**51, **4:**51f
Reef(s), coral, **1:**116–118, **1:**116f, **1:**117f
Reforestation, **5:**57–58
Refuge(s), wildlife, **2:**107–110, **2:**108f,
 2:109t

Regeneration, **4:**92
 defined, **4:**101
Renewable, defined, **1:**77
Renewable energy, defined, **5:**33
Repository, defined, **4:**87
Reptile(s), **1:**27
Resource(s), natural, defined, **3:**86
Resource Conservation and Recovery
 Act (RCRA), **4:**68–69, **4:**77
Respiration, defined, **4:**16
Reverse osmosis, defined, **2:**106
Rhine River watershed, **5:**54f
Rhino(s), **5:**68f
Rhode Island Schools Recycling Club,
 5:115
Rice
 in ancient China, **3:**28
 as food source, **3:**95
Right(s), human
 abuses of, **5:**4
 for indigenous peoples, **5:**6, **5:**6f
 sustainable society and, **5:**5–7
"Ring of Fire," **1:**126
Riparian, defined, **4:**121
River(s), **1:**13, **1:**14f, **1:**102–103, **1:**103t
 environmental concerns of, **1:**105
River systems, **2:**121–123, **2:**122t
Riverside Eco Park, **5:**77
Road(s), during Industrial Revolution,
 3:64–65
Robbins, Ocean, **5:**106–107, **5:**106f
Rock(s), **1:**3, **1:**4f, **1:**4t
 classes of, **1:**3, **1:**4t
 parent, in soil formation, **2:**57, **2:**57f
Rock cycle, **1:**4f
Rock quarrying, in ancient Egypt, **3:**20
Rocky shores, **1:**113–114
Roman Empire
 agriculture in, **3:**24–25
 human impact on, **3:**26–27
Rome, ancient, **3:**14f, **3:**24–27, **3:**25f
 agriculture in, **3:**24–25
 benefits of Romans in, **3:**26
 economy of, **3:**25, **3:**25f
 human impact on, **3:**26–27
 natural resources in, **3:**24–25
 technology of, **3:**26
 trade of, **3:**25, **3:**25f
Roosevelt, Theodore, **4:**117, **5:**115
Rosebud Sioux reservation, wind farm
 on, **5:**19f
Round bacteria, **1:**23f
Roundabout(s), **5:**101–102

Safe Drinking Water Act, **2:**68, **2:**92
 in water pollution treatment, **4:**47
Safety, traffic congestion and, **5:**100–102,
 5:101f
Safety measures, for drinking water,
 2:91–92, **2:**92t
Salinity, defined, **1:**121, **4:**101
Salinization, defined, **1:**99, **2:**106, **3:**17,
 3:17f, **3:**32
Salmon Safe, **5:**78–79
Salt(s), ocean, composition of, **1:**12f
Salt marshes, **1:**110–112, **1:**110f
 environmental concerns of, **1:**112
 ocean water pollution effects on, **4:**52
Salt Road, **3:**24
Saltwater, irrigating crops with, **5:**44
Saltwater intrusion, **4:**46
Saltwater marshes, **1:**14
Saltwater resources, **2:**98–105
Sanctuary, defined, **2:**125, **4:**121, **5:**69
Sand, particle size in, **2:**58, **2:**58f
Sanitary landfills, **4:**59–62, **4:**61f, **4:**62f

Satellite(s)
 Landsat, photograph of Earth's surface
 by, **1:**1, **1:**1f
 space, *Vanguard I,* **5:**25
Savanna(s), **1:**85f, **1:**86–87, **1:**88f
SC Johnson Wax, **5:**74
Scavenger(s)
 defined, **1:**99
 in ecosystem, **1:**43–44, **1:**44f
School(s)
 eco-efficiency in, **5:**77
 sustainable practices in, **5:**115–116,
 5:116f
Scrubber(s), in air pollution control,
 4:11, **4:**12f
Scrubland(s), **1:**93–94, **1:**94f
Sea otters, **1:**30
Seabed ocean turbines, **5:**23, **5:**23f
Seattle Chapter of the National
 Audubon Society, **5:**104f
Seaweed, **5:**48–49, **5:**49f
Secondary succession, **1:**131–132
Sedentism, challenges to, **3:**13–14
Sediment(s), pollution due to, **4:**39–40
Seed plants, **1:**26, **1:**26f
Seed-tree cutting, **5:**56
Selective breeding, defined, **3:**49
Selective cutting, **5:**57, **5:**57f
Serengeti National Park, **4:**117
 lions in, **5:**67f
Serengeti Plains National Park, **1:**86,
 1:86f, **1:**87, **1:**87f
Serf(s), defined, **3:**49
Service(s)
 defined, **3:**86, **5:**84
 environmental impact of, **3:**74–77,
 3:75t, **3:**76f
 production of, **3:**74–77, **3:**75t, **3:**76f
 stages in, **3:**77–78
 impact on natural resources, **3:**78–80,
 3:78t, **3:**79f
Sessile, defined, **1:**121
Seventeen, **5:**108
Sewage treatment plant, in water
 pollution treatment, **4:**46–47, **4:**47f
Shark(s), cracking down on, **5:**47
Shelter-wood harvesting, **5:**56
Shenandoah, **2:**111
Shovel plow, **3:**37f
Sick building syndrome, **4:**12, **4:**13f
 defined, **4:**16
Silent Spring, **5:**114–115
Silicate(s), **2:**65
Silk Road, **3:**28f, **3:**29
Silt, particle size in, **2:**58, **2:**58f
Siltation
 defined, **5:**69
 of waterways, **5:**55
Slash and burn, **1:**74f
 alternatives to, **4:**93
 deforestation due to, **4:**90–92, **4:**91f
Slash-and-burn farming, in America,
 3:38, **3:**39f
Slater, Samuel, in Industrial Revolution
 in America, **3:**53, **3:**54f
Slime molds, **1:**24
Smelting
 of minerals, **2:**72
 stages in, **3:**80
Smog, **4:**7–8
 defined, **4:**1, **4:**16
 industrial, **4:**7–8
 photochemical, **4:**8
Sneed, Cathrine, **5:**105f
Snow leopards, **2:**124f
Snowy egret, **2:**118f

ABOUT THE AUTHORS

JOHN MONGILLO is a noted science writer and educator. He is coauthor of *Encyclopedia of Environmental Science*, and *Environmental Activists*, both available from Greenwood.

PETER MONGILLO has won several awards for his teaching, including School District Teacher of the Year, National Endowment for the Humanities Fellowship Award, and the National Council for Geographic Education Distinguished Teacher Award.